Approaches to Reducing the Use of Forced or Child Labor: Summary of a Workshop on Assessing Practice

John Sislin and Kara Murphy, Rapporteurs

Policy and Global Affairs

NATIONAL RESEARCH COUNCIL
OF THE NATIONAL ACADEMIES

THE NATIONAL ACADEMIES PRESS
Washington, D.C.
www.nap.edu

THE NATIONAL ACADEMIES PRESS 500 Fifth Street, N.W. Washington, D.C. 20001

NOTICE: The project that is the subject of this report was approved by the Governing Board of the National Research Council, whose members are drawn from the councils of the National Academy of Sciences, the National Academy of Engineering, and the Institute of Medicine. The members of the committee responsible for the report were chosen for their special competences and with regard for appropriate balance.

This project was supported by the U.S. Department of Labor, Bureau of International Labor Affairs. Any opinions, findings, conclusions, or recommendations expressed in this publication are those of the author(s) and do not necessarily reflect the views of the organizations or agencies that provided support for the project.

International Standard Book Number-13: 978-0-309-14528-2
International Standard Book Number -10: 0-309-14528-7

Additional copies of this report are available from the National Academies Press, 500 Fifth Street, N.W., Lockbox 285, Washington, D.C. 20055; (800) 624-6242 or (202) 334-3313 (in the Washington metropolitan area); Internet, http://www.nap.edu

Suggested citation: National Research Council. 2009. Approaches to Reducing the Use of Forced or Child Labor. Washington, D.C.: The National Academies Press.

Copyright 2009 by the National Academy of Sciences. All rights reserved.

Printed in the United States of America.

THE NATIONAL ACADEMIES
Advisers to the Nation on Science, Engineering, and Medicine

The **National Academy of Sciences** is a private, nonprofit, self-perpetuating society of distinguished scholars engaged in scientific and engineering research, dedicated to the furtherance of science and technology and to their use for the general welfare. Upon the authority of the charter granted to it by the Congress in 1863, the Academy has a mandate that requires it to advise the federal government on scientific and technical matters. Dr. Ralph J. Cicerone is president of the National Academy of Sciences.

The **National Academy of Engineering** was established in 1964, under the charter of the National Academy of Sciences, as a parallel organization of outstanding engineers. It is autonomous in its administration and in the selection of its members, sharing with the National Academy of Sciences the responsibility for advising the federal government. The National Academy of Engineering also sponsors engineering programs aimed at meeting national needs, encourages education and research, and recognizes the superior achievements of engineers. Dr. Charles M. Vest is president of the National Academy of Engineering.

The **Institute of Medicine** was established in 1970 by the National Academy of Sciences to secure the services of eminent members of appropriate professions in the examination of policy matters pertaining to the health of the public. The Institute acts under the responsibility given to the National Academy of Sciences by its congressional charter to be an adviser to the federal government and, upon its own initiative, to identify issues of medical care, research, and education. Dr. Harvey V. Fineberg is president of the Institute of Medicine.

The **National Research Council** was organized by the National Academy of Sciences in 1916 to associate the broad community of science and technology with the Academy's purposes of furthering knowledge and advising the federal government. Functioning in accordance with general policies determined by the Academy, the Council has become the principal operating agency of both the National Academy of Sciences and the National Academy of Engineering in providing services to the government, the public, and the scientific and engineering communities. The Council is administered jointly by both Academies and the Institute of Medicine. Dr. Ralph J. Cicerone and Dr. Charles M. Vest are chair and vice chair, respectively, of the National Research Council.

www.national-academies.org

Planning Committee for a Workshop on Approaches to Reducing the Use of Forced or Child Labor

Susan Berkowitz, *Chair,* Senior Study Director, Westat

Kevin Bales, Cofounder and President, Free the Slaves

Donna E. Chung, Trade and Labor Compliance Advisor, Sandler, Travis & Rosenberg P.A.

Eric Edmonds, Associate Professor of Economics, Department of Economics, Dartmouth College

Adam B. Greene, Vice President, Labor Affairs & Corporate Responsibility, U.S. Council for International Business

Beryl Levinger, Distinguished Professor of Nonprofit Management, Monterey Institute of International Studies

Dan Viederman, Executive Director, Verité

Staff
Peter Henderson, Board Director
John Sislin, Program Officer
Kara Murphy, Research Associate
Sabrina Hall, Program Associate

PREFACE

In response to provisions of the Trafficking Victims Protection and Reauthorization Act (TVPRA), the U.S Department of Labor's Bureau of International Labor Affairs contracted with the National Research Council (NRC) to organize a two-day workshop on a framework for assessing practices designed to reduce the use of child and forced labor in supply chains that produce goods imported into the United States. To carry out this task, the NRC appointed a planning committee that was charged with developing a draft framework for assessing practice that would be presented and discussed at the workshop which would be summarized in a final report.

The workshop and the draft documents presented at it by the committee were not seen as in any way comprehensive or final products of the committee or of the NRC but rather as a way to start a conversation that would be helpful to DOL in its work. The committee understood that the DOL was beginning a process of producing a compendium and that this workshop was only an initial step, and one of many, that DOL would take in collecting information and organizing the compendium. The committee intended the workshop and this summary report to fit into that broader effort by the DOL, rather than offer any final conclusions or pronouncements.

Indeed, during the course of the workshop, ILAB staff stressed that the workshop was only one step in a longer process, with multiple opportunities for stakeholder involvement. Dr. Charita Castro noted that once the workshop summary was received from the NRC, ILAB would review the material that emerges from the workshop and then solicit additional feedback on the framework and further examples of practice. Moreover, ILAB would provide additional opportunities for input from the business community, NGO counterparts and the broader public.

This summary report of the workshop has been prepared by the rapporteurs as a factual summary of what occurred at the workshop, supplemented by resource material. The planning committee's role was limited to planning and convening the workshop. The statements made are those of the rapporteurs or individual workshop participants and do not necessarily represent the views of all workshop participants, the planning committee, or the National Academies.

This report has been reviewed in draft form by individuals chosen for their diverse perspectives and technical expertise, in accordance with procedures approved by the National Research Council's Report Review Committee. The purpose of this independent review is to provide candid and critical comments that will assist the institution in making its published report as sound as possible and to ensure that the report meets institutional standards for objectivity, evidence, and responsiveness to the study charge. The review comments and draft manuscript remain confidential to protect the integrity of the process.

We wish to thank the following individuals for their review of this report: Chisara Ehiemere, Transfair USA; Adam Greene, U.S. Council for International Business; Ted Moran, Georgetown University; Roger Plant, International Labour Organisation, Switzerland; and Edward Potter, Coca-Cola Company. Although the reviewers listed above have provided many constructive comments and suggestions, they were not asked to endorse the content of the report, nor did they see the final draft of the report before its release.

The review of this report was overseen by Mary Clutter, National Science Foundation (Retired). Appointed by the National Academies, she was responsible for making certain that an independent examination of this report was carried out in accordance with institutional procedures and that all review comments were carefully considered. Responsibility for the final content of this report rests entirely with the authors and the institution.

CONTENTS

1. Introduction   1
    Scope of Child and Forced Labor, 1
    Legislative Context, 4
    Planning the Workshop, 5

2. Scope of the Workshop   9
    Introductory Remarks, 9
    Sponsor Perspectives, 10
    Discussion, 15

3. Assessing the Context of Child and Forced Labor   19
    Problem Identification, 19
    Sectors, 21
    Actors, 23
        Supply Chains, 23
        Partnerships, 25
    Tools, 27
    Key Questions, 29
    Additional Comments, 29

4. Illustrative Business Practices   31
    Introduction, 31
    The Cocoa Sector, 31
        The Cocoa Industry Certification Program, 34
        The Sustainable Tree Crops Program, 35
        The International Cocoa Initiative, 36
    Target, 37
    Levi Strauss and Company, 38
    Fair Labor Association, 43
    Winrock International, 44
    International Labor Rights Forum, 47
    International Labor Organization, 47

5. Criteria   55
    Comments on the Criteria, 58
        General Comments, 58
        Specific Comments on Individual Criteria, 62

6. Wrap-up   65

Appendixes

A.	Committee member biographies	71
B.	Workshop agenda	74
C.	Speaker biographies	77
D.	Participants list	82
E.	Definitions of Child and Forced Labor	84
F.	Illustrative Examples of CSR Practices	91
G.	Submissions by presenters and/or audience members	108
H.	Submissions following the Workshop	121

Boxes

1-1 Definition of Child Labor, 2
1-2 Definition of Forced Labor, 3

4-1 Levi Strauss & Co. Global Sourcing and Operating Guidelines, 40

5-1 ILO Good Practices, 56
5-2 Criteria for Assessing Practices: Draft Proposed at Workshop to Facilitate Discussion, 57

Figures

2-1 Organization of the Bureau of International Labor Affairs at the U.S. Department of Labor, 10

3-1 Perspectives: Forced and Child Labor, 20
3-2 Defining the Challenge/Context in West African Cocoa Production, 22
3-3 Soccer Ball Production and Distribution, 24
3-4 Cocoa Supply Chains in Cote d'Ivoire and Ghana, 25
3-5 Soya Supply Chain in Brazil, 26

4-1 Cocoa Sector Country Certification Model, 34

1
INTRODUCTION

Scope of Child and Forced Labor

Globally, child labor and forced labor are widespread and complex problems. They are conceptually different phenomena, requiring different policy responses, though they may also overlap in practice.

According to the International Labor Organization (ILO), "In 2004 there were 218 million children trapped in child labour, of whom 126 million were in hazardous work."[1] More recently, but employing a different definition, UNICEF estimated that "there are 158 million children under the age of 15 who are trapped in child labour around the world."[2] Most child labor occurs in agriculture (69 percent), as compared with services (22 percent) and industry (9 percent).[3] Child labor was most prevalent in sub-Saharan Africa followed by Asia and the Pacific. According to the ILO, "The vast majority of child labour is found in the informal economy."[4] Box 1-1 provides a definition of child labor from the U.S Department of Labor.[5]

---

[1] ILO. *The End of Child Labour: Within Reach*. Global Report under the follow-up to the ILO Declaration on Fundamental Principles and Rights at Work, International Labour Conference, 95th Session 2006, Report I (B). Geneva: ILO, 2006.

[2] Amy Bennett. "World Day against Child Labour highlights the right of every child to an education." June 12, 2008. Available at: *http://www.unicef.org/protection/index_44437.html* (accessed May 18, 2009).

[3] ILO, *The End of Child Labour: Within Reach*, pp. 7-8. "The agricultural sector comprises activities in agriculture, hunting, forestry and fishing. The industry sector consists of mining and quarrying, manufacturing, construction and public utilities (electricity, gas and water). The services sector includes wholesale and retail trade, restaurants and hotels, transport, storage and communications, finance, insurance, real estate and business services, and community, social and personal services."

[4] Ibid.

[5] The ILO's International Programme on the Elimination of Child Labor (IPEC) provides an alternate definition that should also be considered: "Not all work done by children should be classified as child labour that is to be targeted for elimination. Children's or adolescents' participation in work that does not affect their health and personal development or interfere with their schooling, is generally regarded as being something positive. This includes activities such as helping their parents around the home, assisting in a family business or earning pocket money outside school hours and during school holidays. These kinds of activities contribute to children's development and to the welfare of their families; they provide them with skills and experience, and help to prepare them to be productive members of society during their adult life." See *http://www.ilo.org/ipec/facts/lang--en/index.htm* (accessed September 25, 2009).

> **Box 1-1**
> **Definition of Child Labor**
>
> "'Child labor' under international standards means all work performed by a person below the age of 15. It also includes all work performed by a person below the age of 18 in the following practices: (A) All forms of slavery or practices similar to slavery, such as the sale or trafficking of children, debt bondage and serfdom, or forced or compulsory labor, including forced or compulsory recruitment of children for use in armed conflict; (B) the use, procuring, or offering of a child for prostitution, for the production of pornography or for pornographic purposes; (C) the use, procuring, or offering of a child for illicit activities, in particular for the production and trafficking of drugs; and (D) work which, by its nature or the circumstances in which it is carried out, is likely to harm the health, safety, or morals of children. The work referred to in subparagraph (D) is determined by the laws, regulations, or competent authority of the country involved, after consultation with the organizations of employers and workers concerned, and taking into consideration relevant international standards. This definition will not apply to work specifically authorized by national laws, including work done by children in schools for general, vocational or technical education or in other training institutions, where such work is carried out in accordance with international standards under conditions prescribed by the competent authority, and does not prejudice children's attendance in school or their capacity to benefit from the instruction received."
>
> Department of Labor, Office of the Secretary. *Notice of Procedural Guidelines for the Development and Maintenance of the List of Goods From Countries Produced by Child Labor or Forced Labor; Request for Information.* Federal Register / Vol. 72, No. 247 / Thursday, December 27, 2007, p. 73378.

Forced labor is also a concern.[6] According to the ILO, "Forced labour is present in some form on all continents, in almost all countries, and in every kind of economy."[7] According to the most recent ILO estimates available (2005): "Today, at least 12.3 million people are victims of forced labour worldwide. Of these, 9.8 million are exploited by private agents, including more than 2.4 million in forced labour as a result of human trafficking. Another 2.5 million are forced to work by governments or by rebel military groups."[8] Using a different definition, Free the Slaves has estimated that there are currently 27 million people in slavery.[9]

---

[6] For a general discussion of forced labor, see for example "Forced Labor in the Global Economy: A Report of Discussions," Program on Human Rights and Justice, Center for International Studies, Massachusetts Institute of Technology, May 14, 2005, available at: *http://web.mit.edu/cis/pdf/Forced_Labor.pdf* (accessed September 28, 2009). Kevin Bales. *Disposable People: New Slavery in the Global Economy.* Berkeley, CA: University of California Press, 2004. Kevin Bales. *Understanding Global Slavery: A Reader.* Berkeley, CA: University of California Press, 2005. See also the websites of Free the Slaves at *http://www.freetheslaves.net/Page.aspx?pid=183* (accessed September 28, 2009) and the ILO's Special Action Programme to Combat Forced Labour at *http://www.ilo.org/sapfl/lang--en/index.htm* (accessed September 28, 2009).

[7] ILO. Eradication of Forced Labour—General Survey Concerning the Forced Labour Convention, 1930 (No. 29), and the Abolition of Forced Labour Convention, 1957 (No. 105). Report of the Committee of Experts on the Application of Conventions and Recommendations, Report III (Part Ib), ILC, 96th Session, Geneva, 2007.

[8] ILO, A Global Alliance Against Forced Labour. Global Report under the Follow-up to the ILO Declaration on Fundamental Principles and Rights at Work. ILC, 93rd Session, Report I (B). Geneva: ILO, 2005, p. 10.

> **Box 1-2**
> **Definition of Forced Labor**
>
> "'Forced labor' under international standards means all work or service which is exacted from any person under the menace of any penalty for its nonperformance and for which the worker does not offer himself voluntarily, and includes indentured labor. 'Forced labor' includes work provided or obtained by force, fraud, or coercion, including: (1) By threats of serious harm to, or physical restraint against any person; (2) by means of any scheme, plan, or pattern intended to cause the person to believe that, if the person did not perform such labor or services, that person or another person would suffer serious harm or physical restraint; or (3) by means of the abuse or threatened abuse of law or the legal process. For purposes of this definition, forced labor does not include work specifically authorized by national laws where such work is carried out in accordance with conditions prescribed by the competent authority, including: any work or service required by compulsory military service laws for work of a purely military character; work or service which forms part of the normal civic obligations of the citizens of a fully self-governing country; work or service exacted from any person as a consequence of a conviction in a court of law, provided that the said work or service is carried out under the supervision and control of a public authority and that the said person is not hired to or placed at the disposal of private individuals, companies or associations; work or service required in cases of emergency, such as in the event of war or of a calamity or threatened calamity, fire, flood, famine, earthquake, violent epidemic or epizootic diseases, invasion by animal, insect or vegetable pests, and in general any circumstance that would endanger the existence or the well-being of the whole or part of the population; and minor communal services of a kind which, being performed by the members of the community in the direct interest of the said community, can therefore be considered as normal civic obligations incumbent upon the members of the community, provided that the members of the community or their direct representatives have the right to be consulted in regard to the need for such services."
>
> Department of Labor, Office of the Secretary. *Notice of Procedural Guidelines for the Development and Maintenance of the List of Goods From Countries Produced by Child Labor or Forced Labor; Request for Information.* Federal Register / Vol. 72, No. 247 / Thursday, December 27, 2007, p. 73378.

Child or forced labor may exist in the supply chains of companies and thus in the production of imports to the United States. As one ILO report notes, forced labor can exist in several ways:

> First, there are the widespread problems affecting small industries, sometimes in remote areas, in developing countries. These are long-standing concerns of the largely informal economy, as in the brick kilns or small garment factories of such South Asian countries as India and Pakistan, which are likely to include deeply

---

[9] See Free the Slaves, "What's the Story," available at: *http://www.freetheslaves.net/Page.aspx?pid=301* (accessed May 18, 2009).

embedded practices of bonded labour... Second, there are the industries which appear to be at risk of forced labour practices within individual developing countries, mainly because of the nature of recruitment practices. There is a very clear risk of forced labour through debt bondage, when temporary workers are recruited through informal and unlicensed intermediaries who entice their recruits through the payment of advances, and then make their profits through a series of inflated charges. In Latin America, forced labour has been detected in a range of industries, some of them export oriented... Third, there are the problems facing multinational enterprises (MNEs) which outsource their production to companies operating in developing countries. This may be an extension of the first issue, given that the goods widely produced under forced and child labour conditions in the small garment and other factories in the developing countries can penetrate the supply chain of MNEs... Fourth, there are the potential problems facing all companies, in developed and developing countries alike, which engage contract labour through different kinds of employment or recruitment agencies."[10]

Child labor could be viewed similarly. There are also many media reports of examples of child or forced labor occurring in specific situations.

Legislative Context

The federal government has enacted laws to reduce the use of child and forced labor in the production of goods consumed in the United States.[11] The Trafficking Victims Protection Act of 2000 (TVPA) had as its purpose: "To combat trafficking in persons, a contemporary manifestation of slavery whose victims are predominantly women and children, to ensure just and effective punishment of traffickers, and to protect their victims."[12] The Act was reauthorized in 2003, 2005, and 2008.

The 2003 reauthorization established a research agenda on domestic and international trafficking in persons, including such topics as the economic causes and consequences of trafficking in persons; the effectiveness of programs and initiatives funded or administered by federal agencies to prevent trafficking in persons and to protect and assist victims of trafficking; and the interrelationship between trafficking in persons and global health risks.

The 2005 reauthorization added a section to the Act entitled "Additional activities to monitor and combat forced labor and child labor." Specifically:

- Section 105(b)(1) of the Act directed the Secretary of Labor, acting through the Bureau of International Labor Affairs, to "carry out additional activities to monitor and combat forced labor and child labor in foreign countries."

- Section 105(b)(2) listed these activities as:
    (A) Monitor the use of forced labor and child labor in violation of international standards;

---

[10] ILO, The Cost of Coercion. Global Report under the Follow-up to the ILO Declaration on Fundamental Principles and Rights at Work. ILC, 98th Session, Report I(B). Geneva: ILO, 2009, pp. 51-52.
[11] Several examples are found in the presentation by U.S. Department of Labor staff summarized in the next chapter.
[12] Public Law 106–386, October 28, 2000.

(B) Provide information regarding trafficking in persons for the purpose of forced labor to the Office to Monitor and Combat Trafficking of the Department of State for inclusion in [the] trafficking in persons report required by Section 110(b) of the Trafficking Victims Protection Act of 2000 (22 U.S.C. 7107(b));
(C) Develop and make available to the public a list of goods from countries that the Bureau of International Labor Affairs has reason to believe are produced by forced labor or child labor in violation of international standards;
(D) Work with persons who are involved in the production of goods on the list described in subparagraph (C) to create a standard set of practices that will reduce the likelihood that such persons will produce goods using the labor described in such subparagraph; and
(E) to consult with other departments and agencies of the United States Government to reduce forced and child labor internationally and ensure that products made by forced labor and child labor in violation of international standards are not imported into the United States.

In the 2008 reauthorization, Section 110 of the Law requires the Department of Labor to produce a report that details a list of goods produced with child or forced labor:

(a) Final Report; Public Availability of List; not later than January 15, 2010, the Secretary of Labor shall
    (1) submit to the appropriate congressional committees a final report that
        (A) describes the implementation of section 105(b) of the Trafficking Victims Protection Reauthorization Act of 2005 (22 U.S.C. 7103(b)); and
        (B) includes an initial list of goods described in paragraph (2)(C) of such section; and
    (2) make the list of goods described in paragraph (1)(B) available to the public.[13]

Planning the Workshop

Motivated by Section 105(b)(2)(D) in the 2005 reauthorization, which directed the Bureau of International Labor Affairs (ILAB) to create a standard set of practices that will reduce the likelihood that such persons will produce goods using child and forced labor, ILAB contracted with the National Research Council (NRC) to organize a workshop that would take the first steps in developing a framework for organizing a standard set of practices.

The NRC organized a planning committee (Appendix A) for this workshop that was guided by the following charge:

An ad hoc committee will convene a public workshop on practices to reduce the use of forced or child labor in the production of goods. In preparation for the workshop, which will feature invited presentations and discussions, the committee will oversee the collection of illustrative examples of such practices. Using this and other information provided at the workshop, participants will discuss the elements of a possible framework for identifying and organizing a standard set of practices that will reduce the likelihood that persons will use forced labor or child labor to produce goods, with a

---

[13] Public Law No: 110-457

focus on business and governmental practices. An individually-authored summary of the workshop will be produced, including examples of some of the business and government practices discussed.

Through one meeting held in Washington, DC, and several teleconferences, the planning committee developed an agenda for the workshop, including suggested presenters and invited audience members, and a draft framework designed to stimulate discussion.

The planning committee conceptualized the draft framework for identifying and organizing a standard set of practices as consisting of two parts that were summarized in two brief documents made available at the workshop: a description of the context in which such practices could be made (and which could affect those practices), and a list of criteria that could be used to filter from among the many business practices those that were particularly noteworthy, considering partnerships with other businesses, associations, nongovernmental organizations (NGOs), and governments.

The committee believed that understanding the context of a practice is important. The criteria used to assess a practice may vary with the type of practice, the sector, or the national, cultural, or socioeconomic environment, suggesting many different criteria would be needed. However, for purposes of drafting a set of criteria that could stimulate discussion, the planning committee developed a parsimonious set of criteria that could be used to judge many different types of business practices. These draft criteria are presented in chapter five of the report.

The purpose of the draft framework, incorporating a discussion of context and criteria, was to stimulate a debate on what makes good or effective practices. As the ILO has noted:

> Good practices provide a means of being able to learn from and to apply experiences of others. Otherwise, one may devote considerable effort in "reinventing the wheel" or in repeating mistakes that others already have made. Good practices can be used most appropriately to stimulate thinking and to suggest ideas for consideration. It is not expected that good practices necessarily should be copied from one setting to another. The context can vary across settings, and thus even highly successful interventions may not "travel" well. At the least, however, these can provide "food for thought" and ideas about possible adaptations. The more that a similar approach has been tried and shown to work in multiple and varied settings, the more likely that it might also apply in some respect elsewhere as well.[14]

The committee, in planning the workshop, did not see the framework as necessarily identifying "best practices," which seemed premature at this juncture, but rather "illustrative practices." Further, the committee took a broad view of what could be included in the notion of practices. Practices "can include policy, planning and research activities, legislation, programmes and projects, as well as 'on-the-ground' delivery of programmes."[15] Examples of practices include: awareness-raising activities, policy development, implementation, monitoring, and forging partnerships.[16]

---

[14] B. Perrin, Combating child labour: sample good practices guidelines. Geneva: ILO, October 2003. p. 1
[15] Ibid., p. 2.
[16] IFC, Addressing child labor in the workplace and supply chain, good practice note, June 2002, Number 1, available at: *http://www.ifc.org/ifcext/enviro.nsf/AttachmentsByTitle/p_ChildLabor/$FILE/ChildLabor.pdf*. Accessed May 28, 2009.

In developing the workshop agenda, however, the committee did focus on practices that businesses could adopt to prevent, reduce, or remediate instances of child or forced labor. While businesses should certainly partner with other actors as appropriate, the committee did not seek to include important practices that were primarily taken by others, such as awareness-raising campaigns or litigation by NGOs; guidance offered by the ILO; labor inspections; or national legislation by governments. The activities of other actors, such as NGOs, are often critical to addressing child and forced labor and the relative responsibility of these actors are today the subject of intense debate, but they were outside the scope of this particular project.

The workshop and the draft documents presented at the event by the committee were not seen as in any way comprehensive or final products of the committee or the NRC but rather as a way to start a conversation that would be helpful to the U.S. Department of Labor (DOL) in its work. The committee understood that DOL was beginning a process of collecting information and organizing the compendium of practices and that this workshop was only an initial step in that process. The committee intended the workshop and this summary report to fit into that broader effort, rather than offer any final conclusions or pronouncements.

The committee felt that since committee members, presenters, international labor standards scholars, and other audience members present a range of different and sometimes conflicting ideas about what works, how to organize practices, and other issues it was premature to offer a single approach to identifying more effective business practices.

The planning committee also felt that with experts on all sides present, there would be a lively discussion of the draft criteria. Presenters with experience in and divergent views on the topic of corporate social responsibility and child or forced labor were invited to speak to the draft criteria. They were asked either to explore the usefulness of the criteria by presenting a single case study of a business practice designed to reduce the use of child or forced labor in the supply chain or to explore more deeply the nuances, challenges, and limitations of one element of the criteria by comparing multiple case studies. To facilitate a greater exchange of information, invitations were sent to potential audience members both to attend the workshop and to submit comments or criticisms of a set of draft criteria regardless of whether the individuals were able to attend. Comments received are reproduced in Appendix G.

The workshop was held on May 11-12, 2009. (Please see Appendix B for the final agenda; Appendix C for speaker biographies; and Appendix D for a list of participants.) In the following pages a summary of the presentations and discussions that took place is presented. Chapter 2 focuses on the scope of the workshop and includes two presentations: welcoming remarks by the planning committee chair, Susan Berkowitz, and a presentation by staff of the DOL on the policy context of the workshop. In Chapter 3 the first half of the framework is discussed through a presentation and discussion of the importance and nature of the context in which practices are undertaken. Chapter 4 illustrates a series of business practices (both undertaken by businesses and businesses in broader partnerships). Chapter 5 presents the draft criteria—the second half of the framework—and provides extensive comment on how DOL should move forward in its development of the criteria the agency will use in its work. Because many presenters and audience members responded to the different presentations throughout the workshop—sometimes offering similar themes or critiques or asking similar questions, the material in the workshop has been grouped into these three main categories—context, business practices, and draft criteria—rather than summarized chronologically as they occurred during the workshop. Chapter 6, the concluding chapter, presents a "wrap up" of the workshop in which planning committee members offered their thoughts on the main themes of the workshop.

2
SCOPE OF THE WORKSHOP

Introductory Remarks

Susan Berkowitz, chair of the planning committee, opened the workshop with several focusing comments. After first welcoming the speakers and guests, Dr. Berkowitz elaborated on the purpose of the workshop, which was to discuss a set of criteria for identifying good practices to reduce the use of child or forced labor. She reminded the audience that the criteria to be presented are a starting point for discussion and by no means final. Dr. Berkowitz noted that the workshop was set up to look at the criteria by "critically assessing them, thinking about where there may be gaps, where we may want to reorganize them." She indicated that the goal of the workshop is "to have tested these criteria against the experience of people who are here either as presenters or in the audience who have deep knowledge of this area because what we would like to emerge from [the workshop] is to flesh out these criteria, to test them against your collective experience, to revise them, and to have them emerge strengthened so that we can go forth with some sense of a set of criteria that can be used across different domains."

Dr. Berkowitz's noted the importance of focusing on good rather than best practices. She argued that the draft criteria were geared more to finding good practices. A limitation in the criteria, in her opinion, was that we were not "quite there in terms of the ability to define a set of practices that would necessarily work across contexts, recognizing that this is an area where context is very important." Dr. Berkowitz continued by noting that context matters. She stated it was essential to think about the importance of context in which business practices operate and to use some flexibility in the application of criteria, arguing that "we always need to be sensitive to the particular context in which any of these criteria would be utilized to identify, help ameliorate, modify, or respond to either child or forced labor. That is why the language is not one of best practices but of good practices or effective practices."

Dr. Berkowitz concluded her welcoming remarks by inviting participation from everyone regardless of whether they were a presenter or attending the workshop as an interested audience member. She downplayed the distinction between these two groups, in particular to facilitate a stronger set of criteria that has a certain buy-in, in essence, from the community. She noted that "obviously in the context of the current economic crises, there is all the more reason to be concerned about these kinds of questions [of child and forced labor]."

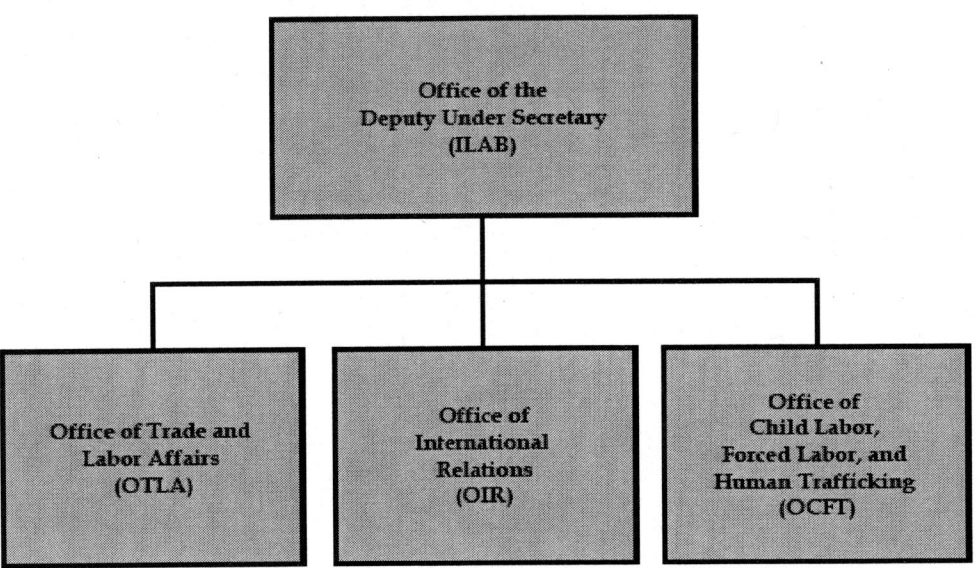

FIGURE 2-1 Organization of the Bureau of International Labor Affairs at the U.S. Department of Labor.
Source: Rachel Rigby, U.S. Department of Labor, Workshop Presentation, May 11, 2009.

Sponsor Perspectives

Ms. Rachel Rigby and Dr. Charita Castro of the Office of Child Labor, Forced Labor, and Human Trafficking (OCFT), Bureau of International Labor Affairs (ILAB), U.S. Department of Labor presented the perspectives of the sponsors at the workshop.

Ms. Rigby presented the organization and mission of ILAB as well as the policy or legislative context for this endeavor. As shown in Figure 2-1, ILAB is composed of an Office of the Deputy Under Secretary comprising senior officials and three operating offices: the Office of Trade and Labor Affairs; the Office of International Relations; and the Office of Child Labor, Forced Labor, and Human Trafficking.

The mission of OCFT is to support the labor and foreign policy objectives of the President and the Secretary of Labor, meet congressional mandates, and perform public outreach by promoting the elimination of the worst forms of child labor and increasing knowledge and information on child labor, forced labor, and human trafficking.[1] The Office's portfolio consists of three areas: (1) technical assistance, the grant funding that the Office provides to support international projects designed to eliminate child labor, forced labor, and human trafficking; (2) research and policy; and (3) awareness-raising, helping to educate people in the United States

---

[1] More information about the Office can be found at: *http://www.dol.gov/ilab/programs/ocft/*.

SCOPE OF THE WORKSHOP

and abroad on international child labor, forced labor, and human trafficking issues.

The Office was founded in 1993 at the request of Congress to research and report on international child labor issues. At that time the Office was focused on child labor and was primarily focused on research. In 1995 the Office began receiving funds for its first technical cooperation projects. Those grants went to the International Labor Organization's International Program on the Elimination of Child Labor.[2] In 2001 Congress appropriated additional funds that were used in the Office's Child Labor Education Initiative[3] to provide grants to organizations that remove or prevent children from the worst forms of child labor through educational programs on the theory that education is the best path toward the reduction and elimination of child labor. The Office has continued to receive increasing funding for its education initiative since 2001.

In 2005 the office received a new mandate under the Trafficking Victims Protection Reauthorization Act (TVPRA) to monitor and report on forced labor and child labor. Ms. Rigby noted that the Office had already been monitoring and reporting on child labor as well as forced labor as it pertained to children all along; the TVPRA broadened the Office's mandate to include the issue of forced labor among adults.

Ms. Rigby then presented a selection of research initiatives currently being funded[4]:

- The current project, through the National Academies, for the development of criteria for the standard set of practices under the TVPRA;
- ILO-IPEC for research that includes national child labor surveys, child labor data collection, methodological developments, baseline surveys, impact assessment, and thematic research;[5]
- Understanding Children's Work Project, a partnership between the World Bank, the ILO, and UNICEF, that conducts work on impact evaluation, indicator development and country-level research;[6]
- Macro International, which conducts research in India, Pakistan and Nepal on children working in the carpet industry—a three year research project;[7]
- Verité, which is undertaking research on forced labor in eight countries in the production of specific goods; and
- Tulane University, which oversees public and private initiatives to eliminate the worst forms of child labor in the cocoa sector in Côte d'Ivoire and Ghana.[8]

---

[2] According to Ms. Rigby, ILO-IPEC continues to be a major partner for her office.
[3] More information on the Initiative can be found at: *http://www.dol.gov/ilab/programs/iclp/education/main.htm* (accessed September 28, 2009). Goals of the program are: (1) Raise awareness of the importance of education for all children and mobilize a wide array of actors to improve and expand education infrastructures; (2) Develop formal and transitional education systems that encourage working children and those at risk of working to attend school; (3) Strengthen national institutions and policies on education and child labor; and (4) Ensure the long-term sustainability of these efforts.
[4] ILO-IPEC funding is tens of millions per year. See "U.S. Department of Labor awards more than $58 million to eliminate exploitive child labor around the world." Reuters, October 1, 2008, and OCFT Project Status at *http://www.dol.gov/ilab/programs/ocft/project-africa.htm* (accessed September 28, 2009) and *http://www.dol.gov/ilab/programs/ocft/project-asia.htm* (accessed September 28, 2009).
[5] For a discussion of ILAB funding to ILO-IPEC from 1995 to 2006, see: *http://www.dol.gov/ILAB/programs/iclp/iloipec/main.htm* (accessed September 28, 2009).
[6] For more information on the program, see: *http://www.ucw-project.org/* (accessed September 28, 2009).
[7] In March 2009 Macro International joined with ICF International and is now called ICF Macro.
[8] See the project's Web site at *http://www.childlabor-payson.org/* (accessed September 28, 2009).

Ms. Rigby then discussed the Office's legislative mandates on child labor, forced labor, and human trafficking, which form an important policy context for this project, noting that the Office has several current initiatives that are specifically mandated by U.S. law.

Executive Order 13126 of 1999, Prohibition of Acquisition of Products Produced by Forced or Indentured Child Labor.[9] The intention of this E.O. was to prevent federal agencies from purchasing products made with forced or indentured child labor; Ms. Rigby emphasized that the focus was on *forced or indentured child labor* and does not include adults. The Office developed and published a list of products in 2001 that it felt met this definition and produced procedural guidelines to guide our work on that list.[10]

At the same time, the General Services Administration published a federal acquisition regulation final rule as part of the overall U.S. government procurement regulations that require federal contractors that supply products on the Office's list to certify that they have made a good faith effort to ensure that these products were not made with forced or indentured child labor.[11] Ms. Rigby noted that the Office's list can be updated in response to public submissions or as a result of DOL's own independent research.

The Trade and Development Act of 2000. The Trade Act of 1974 established the U.S. generalized system of preferences (GSP), a trade preference program primarily aimed at developing countries.[12] Subsequent amendments to the Trade Act required annual reports to Congress on the status of internationally recognized worker rights, including country efforts to uphold their commitments to combat the worst forms of child labor. The 2000 Trade and Development Act was passed, which called on DOL to report annually to Congress with specific information on beneficiary countries and how they were upholding their commitment to combat the worst forms of child labor. The report covers countries receiving GSP benefits and benefits under other trade preference programs: the African Growth and Opportunity Act (AGOA), the Andean Trade Preference Act (ATPA), and the Caribbean Basin Trade Partnership Act (CBTPA).[13] The report covers about 140 countries, and varies each year depending on what

---

[9] For the text of the Executive Order see: *http://frwebgate.access.gpo.gov/cgi-bin/getdoc.cgi?dbname=1999_register&docid=99-15491-filed.pdf* (accessed September 28, 2009).

[10] Details on the 2001 list are found at: *http://frwebgate.access.gpo.gov/cgi-bin/getdoc.cgi?dbname=2001_register&docid=01-953-filed.pdf* (accessed September 28, 2009). For the text of the Procedural Guidelines, please see: *http://frwebgate.access.gpo.gov/cgi-bin/getdoc.cgi?dbname=2001_register&docid=01-952-filed.pdf* (accessed September 28, 2009).

[11] For the text of the Final Rule see: *http://frwebgate.access.gpo.gov/cgi-bin/getdoc.cgi?dbname=2001_register&docid=01-1503-filed.pdf* (accessed September 28, 2009).

[12] For text of the Act, see Title 19, Chapter 12, available at: *http://www4.law.cornell.edu/uscode/19/ch12.html* (accessed September 28, 2009).

[13] For an example, see: The Department of Labor's 2007 Findings on the Worst Forms of Child Labor available at: *http://www.dol.gov/ilab/programs/ocft/PDF/2007OCFTreport.pdf* (accessed September 28, 2009). Links to all reports are available at: *http://www.dol.gov/ilab/media/reports/iclp/main.htm* (accessed September 28, 2009).

countries are actually receiving those benefits at any given time. It includes various sections for each country. There is data on working children and the nature of child labor in the country, information on minimum age of work laws and legislation that pertains to the worst forms of child labor in the country such as Penal Codes, information on child labor law enforcement, and overview of government policies and program to combat child labor.

The TVPRA of 2005.[14] The TVPRA of 2005, Section 105, contained five new mandates for the DOL. Part A calls for the DOL to monitor the use of forced labor and child labor in violation of international standards. As Ms. Rigby stated, the Office has been monitoring the use of child labor since its founding in 1993; the change was the additional focus on monitoring forced labor of adults. An outcome of this mandate was that the Office's research portfolio expanded.

Part B requires DOL to provide information regarding trafficking in persons for the purpose of forced labor to the U.S. State Department's Office to Monitor and Combat Trafficking in Persons (G/TIP). This, Ms. Rigby noted, was not a new mandate, and in fact the two offices had been working closely and sharing information since the G/TIP was founded in 2001. G/TIP publishes an annual report *Trafficking in Persons,*[15] which contains information supplied by DOL, and DOL cites these reports in its annual reports.

Part C requires ILAB to develop and make available to the public a list of goods from countries that ILAB has reason to believe are produced by forced labor or child labor in violation of international standards. Ms. Rigby noted that this area of the mandate had perhaps received the most attention thus far. She noted that ILAB was in the process of producing this list (due in early 2010). ILAB started by setting out procedural guidelines that were published in draft form in October 2007. Public input was then solicited and considered in the final guidelines published in December 2007.[16] The guidelines lay out the process by which ILAB will make decisions about which goods should go on that list, the criteria that inform those decisions as well as procedural issues on the development of the list, what will happen after the list is published, and how can it be modified. Ms. Rigby highlighted a key element of these procedural guidelines, which was to point out that they contain the definitions that ILAB uses for forced labor and child labor in the development of this list, as well as other key terms such as what constitutes a "good."[17] Ms. Rigby noted that the definitions are based on both international standards and U.S. law.[18] The definition of child labor used is based on ILO Conventions 138 and 182. Convention 138 stipulates that the minimum age for work should be no less than 15 or 14 in certain less developed country contexts and Convention 182 lays out the internationally recognized worst forms of child labor that no child under age 18 should be involved in, including work that could be harmful to the health, safety, or morals of children. For forced labor, guidance is taken from ILO Convention 29.

After the procedural guidelines were published, the public was asked to share with ILAB any information on forced labor or child labor and the production of goods globally. The request was framed broadly to allow for a wide scope of comments. The request is essentially open-

---

[14] Reauthorized in 2008.
[15] For an example see 2008 Trafficking in Persons Report available at: *http://www.state.gov/g/tip/rls/tiprpt/2008/* (accessed September 28, 2009).
[16] To read the draft procedural guidelines, see: *http://edocket.access.gpo.gov/2007/pdf/E7-19310.pdf* (accessed September 28, 2009). To see the final guidelines, see: *http://edocket.access.gpo.gov/2007/pdf/E7-25036.pdf* (accessed September 28, 2009).
[17] The definitions are presented in the first chapter of this report.
[18] For a further discussion of the definitions, in particular as they relate to ILO Conventions, see Appendix E.

ended. ILAB also held a public hearing in April 2008. Six witnesses testified, about 100 individuals attended, and the event was broadcast on C-SPAN.[19] ILAB intends to publish that list by January 15, 2010.

Part D of the TVPRA called on DOL to work with persons who are involved in the production of listed goods described in subparagraph C, to create a standard set of practices that will reduce the likelihood that such persons will produce goods using the labor described in that paragraph (i.e., forced labor and child labor). As Ms. Rigby noted, this is the part of the mandate that gave rise to this workshop.

Part E calls on the DOL to consult with other departments and agencies of the U.S. government to reduce forced and child labor internationally and ensure that products made by forced labor and child labor in violation of international standards are not imported into the United States. As Ms. Rigby noted, the DOL does not have enforcement capacity; rather DOL shares information with other departments and agencies of the U.S. government who do have mandates that relate to the importation of goods. The DOL shares information about the research that it has gathered, which these other entities can use to make decisions about trade policy or importation, for example.

The Food Conservation and Energy Act of 2008. Known as the Farm Bill, this legislation mandated that the U.S. Department of Agriculture establish a consultative group to eliminate the use of child labor and forced labor in imported agricultural products. The DOL has a specifically mandated role in the legislation that is part of that group. The group is composed of 13 members, one of which will be the DOL's Deputy Under Secretary for International Affairs and 12 other members drawn from governmental, nongovernmental, and private sectors.[20] The Farm Bill requires that the group develop recommendations relating to guidelines to reduce the likelihood that agricultural products or commodities imported into the United States are produced with the use of forced labor and child labor.[21] The Bill also makes specific reference to the TVPRA, and Ms. Rigby highlighted the close tie between the list that DOL is producing under the TVPRA, which may or may not contain agricultural goods, and the guidelines that will be produced by this group to reduce the likelihood that agricultural goods will be produced using forced labor and child labor.

---

[19] For a record of the hearing, please see: *http://www.dol.gov/ilab/programs/ocft/pdf/20080423g.pdf* (accessed September 28, 2009).

[20] Further details provided by DOL can be found at: *http://www.dol.gov/ILAB/programs/ocft/fcea.htm* (accessed September 28, 2009).

[21] According to the Federal Register Notice (January 21, 2009, 74, No. 12, Page 3546-3547): "The Consultative Group will develop recommendations relating to a standard set of practices for independent, third-party monitoring and verification for the production, processing, and distribution of agricultural products or commodities to reduce the likelihood that agricultural products or commodities imported into the United States are produced with the use of forced labor or child labor. Recommendations developed by the Consultative Group will be submitted to the Secretary of Agriculture by June 18, 2010. Thereafter, the Consultative Group will continue to advise the Secretary as necessary."

Dr. Charita Castro provided further perspectives from the sponsor on the purpose of the workshop, stating that the impetus for it lay in the TVPRA of 2005,[22] which called on DOL to work with producers to create a standard set of practices to reduce the likelihood that goods will be produced using forced labor or child labor globally.[23] She stated that this workshop was a good venue to bring together experts working in the field and that those presenters represent a broad range of experts from the field of child labor, forced labor, corporate social responsibility, and best practices theory.

Dr. Castro said that OCFT has been funding projects since 1995, including projects on forced child labor in over 75 countries.[24] These funds have helped the Office gain a great deal of knowledge and expertise among their grantees, particularly implementing agencies such as the ILO (including the International Programme on the Elimination of Child Labour (ILO-IPEC)), and other grantees, some of whom were at the workshop. She noted that these organizations have worked on various interventions and approaches to address child labor and forced labor, working on direct interventions to withdraw and prevent individuals from this exploitation. They have also worked closely with governments and the private sector. Thus, she concluded, that in examining good practices, DOL already had a strong knowledge base to start from. However, she noted that there are also practices by the private sector, public-private partnerships, and other stakeholders that DOL may not be aware of or that have not been shared with the broader public.

Given the mandate of the TVPRA and DOL's state of knowledge, Dr. Castro posed the question: how can the DOL best identify practices that have been most effective so that these can be disseminated and replicated? The DOL approached the National Research Council to help OCFT create a framework that could be used to evaluate practices and identify the most effective ones or the ones that are emerging in the fields that can be considered good practices.

Dr. Castro noted that the framework was meant to be a tool to evaluate company efforts, but also noted its potential for use in assessing efforts by governments, public-private partnerships, and others to combat child labor and forced labor in global supply chains.[25] Dr. Castro hoped that that over the next two days DOL would get effective feedback as well as a thoughtful discussion about what would be useful criteria. Dr. Castro stressed that this was only one step in a longer process, with multiple opportunities for stakeholder involvement. She noted that once the workshop summary was received from the National Academies, DOL would provide additional opportunities for input from the business community, NGO counterparts, and the broader public. DOL will review the material that emerges from the workshop and then solicit additional feedback on the framework and further examples of practices that go beyond this meeting.

Discussion

At this point, the floor was opened to comments and questions from the audience. One participant asked for clarification of the starting point for the workshop and whether there was a

---

[22] The TVPRA is discussed in more detail in Chapter 1.
[23] Dr. Castro noted in her presentation that the DOL follows international standards and in particular ILO conventions in defining "child labor" and "forced labor." The actual definitions were not presented in the speech, but can be found in the boxes in the first chapter.
[24] See for instance U.S. Department of Labor, *Faces of Change: Highlights of U.S. Department of Labor Efforts to Combat International Child Labor*, Second Edition. Washington, D.C.: USDOL, 2008.
[25] Dr. Castro noted that for purposes of the workshop, that DOL was not distinguishing between goods that are imported into the United States or goods used for domestic consumption. Both types were considered.

set of assumptions guiding the endeavor. Ms. Rigby noted in response that there are many stakeholders engaged in combating forced and child labor and that through this workshop DOL sought to narrow their focus to private-sector practices. She noted that ILAB was already familiar with many NGO activities, so the workshop was meant to complement what they already knew by focusing on the private sector and public-private partnerships more directly. Dr. Castro added that the focus was on the production of goods, not service industries.

A participant noted that one piece of legislation that seemed to be missing from the discussion provided by Ms. Rigby was the Smoot-Hawley Tariff Act. The questioner suggested that the TVPRA, in 105(b)(e), pretty much replicates what was in the early tariff requirements about the exclusion of any good produced with forced labor and then later into child labor, but it also included labor and products produced in prisons. The questioner asked whether the TVPRA took account of the potential use of forced labor in the flow of products from prison-based systems. Ms. Rigby responded that the reason the Smoot-Hawley Act was not covered in her presentation was that it is a law enforced only by the Department of Homeland Security (DHS) Immigration and Customs Enforcement (ICE). According to the Act, ICE is responsible for prohibiting the importation of goods made by prison labor or forced labor under penal sanctions, and that includes of children.[26] Ms. Rigby pointed out that a key definitional issue was whether prison labor is forced labor. Based on the ILO Convention 29 definition of forced labor, some prison labor is included, but only under what is defined by the ILO. Not all prison labor is forced labor.

A third question focused on Executive Order 13126. The audience member noted that under the executive order contractors are supposed to provide a good-faith examination of their supply chains to determine whether there is forced labor or child labor.[27] The question for DOL

---

[26] One participant then added, "The original act says all goods, wares, articles and merchandise mined, produced, or manufactured wholly, or in part, in any foreign country by convict labor and or forced labor and/or indentured labor under penal sanctions. That was the 1930 law and then it was later amended to include child labor." The full text is: "All goods, wares, articles, and merchandise mined, produced, or manufactured wholly or in part in any foreign country by convict labor or/and forced labor or/and indentured labor under penal sanctions shall not be entitled to entry at any of the ports of the United States, and the importation thereof is hereby prohibited, and the Secretary of the Treasury is authorized and directed to prescribe such regulations as may be necessary for the enforcement of this provision. The provisions of this section relating to goods, wares, articles, and merchandise mined, produced, or manufactured by forced labor or/and indentured labor, shall take effect on January 1, 1932; but in no case shall such provisions be applicable to goods, wares, articles, or merchandise so mined, produced, or manufactured which are not mined, produced, or manufactured in such quantities in the United States as to meet the consumptive demands of the United States. "Forced labor," as herein used, shall mean all work or service which is exacted from any person under the menace of any penalty for its nonperformance and for which the worker does not offer himself voluntarily. For purposes of this section, the term "forced labor or/and indentured labor" includes forced or indentured child labor."

[27] According to the language of the Executive Order, "Each solicitation of offers for a contract for the procurement of a product included on the list published under section 2 of this order shall include the following provisions: (1) A provision that requires the contractor to certify to the contracting officer that the contractor or, in the case of an incorporated contractor, a responsible official of the contractor has made a good faith effort to determine whether forced or indentured child labor was used to mine, produce, or manufacture any product furnished under the contract

was whether "it has supplied what it believes for the past 10 years to have been a definition of good faith efforts. What has been done? What is the standard or definition of those good faith efforts relating to monitoring and certification?" Ms. Rigby clarified that there are two parts to the Executive Order. One part places primary responsibility on DOL for the "research and production of a list of the goods that DOL believes are habitually made with those forms of labor." She noted that when contractors undertake that good-faith effort, they provide the certifications to a different agency.

A fourth question dealt with the roles of the DOL in relation to the State Department and DHS. The questioner was concerned that the agencies might be operating in silos and were not cooperating as much as they could. Dr. Castro responded that the DOL had a number of ways to gather information, including independent contractors collecting data, including primary data from other countries; research activities of the program managers at the OCFT (as well as hearing from grantees); and information provided by the State Department, DOL's "eyes and ears on the ground." She noted that DOL has a variety of ways to gather information on the ground even though the DOL does not have field offices. Dr. Castro also noted that DOL worked collaboratively with ICE.

A fifth question related to part D of the TVPRA, calling for DOL to work with producers of goods on the list developed to create a set of standards to reduce the likelihood that such persons will produce goods using child labor or forced labor. The questioner asked how DOL intended to interpret that, specifically whether the focus would be on those persons who are engaged in the use of child labor in the production of those goods or those who are not but who might be interested in setting standards for an industry or a sector so as to discourage others from doing so? Dr. Castro responded that they try to engage multiple stakeholders, but "as far as specifics, we are still in discussions about this at DOL."

Following up on that question, a participant noted that the draft criteria provided to workshop participants ahead of time (and discussed in Chapter 5) did not seem to address incentives (including negative incentives) to encourage compliance, nor did it touch upon remediation so that the system or process would not just be regulatory but also encourage and perhaps even help with the changes. The participant wondered if those topics were on the table, to which Dr. Castro responded that they were. One audience member added that it might also be helpful to consider the impact on the wider population and the wider economy, or the wider society of operations that intervene in forced or child labor. One example would be the use of an economic boycott against a country where "we hammer down on the government which then leads to economic deprivation of significant parts of the population."

The discussion concluded with comments regarding labor practices in the United States. One participant asked: "Considering that we have extensive privately owned prisons in the United States and quite a number of them produce under contract to the U.S. government, I am wondering how that is going to fit into these standards or the criteria." Another audience member noted "the fact that the TVPRA talks about imported agricultural products but of course we have very significant problems with forced labor in domestic agriculture products." We have to talk about wages and hours. We also have to talk about labor inspection and we have to hope

---

and that, on the basis of those efforts, the contractor is unaware of any such use of child labor; and (2) A provision that obligates the contractor to cooperate fully in providing reasonable access to the contractor's records, documents, persons, or premises if reasonably requested by authorized officials of the contracting agency, the Department of the Treasury, or the Department of Justice, for the purpose of determining whether forced or indentured child labor was used to mine, produce, or manufacture any product furnished under the contract."

for an increase in labor inspectors in the United States that might even someday match, say, the number in Ghana per capita." The first audience member concluded by noting that when individuals work on these issues overseas, they are challenged by individuals in foreign countries who point out flaws they see in the United States. The commenters thought that DOL should take this into account.

In the exchange that followed Ms. Rigby reiterated that DOL was following international labor standards, regardless of U.S. practice or practice of any individual country. Dr. Castro noted that DOL receives similar questions when its staff travel overseas. She noted that for purposes of this project, the focus is on the international context, though that is not to say the domestic situation is not important.

3
ASSESSING THE CONTEXT
OF CHILD AND FORCED LABOR

Business practices designed to address the incidence of child or forced labor in a supply chain that leads to the production of imports to the United States operate in a complex social, cultural, economic, and governmental context. This section summarizes that section of Dr. Berkowitz's presentation that provided a description of the context in which business practices must be implemented and the criteria for assessing practice must be applied. Integrated within this summary is commentary by presenters and audience members who raised issues relating to context.

Dr. Berkowitz began by describing how the planning committee's notion of the framework had two components—the context and the criteria—that were connected but which would be focused on separately. Dr. Berkowitz then discussed the heuristic shown in Figure 3-1 that the planning committee developed as a way of starting a conversation at the workshop about the context. As Dr. Berkowitz discussed, context can be thought of as consisting of several dimensions: problem identification, sectors, actors, tools, and key questions, each of which is described and discussed in turn below.

Problem Identification

The committee argued that the first task is to identify the problem, that is, child and forced labor. These concepts were defined in Chapter 1, but participants offered three views of the problem that merit further discussion.

First, it was important to separate child and forced labor as distinct problems with different root causes. "It seems to me when we are talking about child labor, we have to take a look at the role that poverty plays in that. It is very different from forced labor. What the alternatives are for children, what the alternatives are for families in terms of surviving and what the alternatives are in some of the situations around the world where it is cultural practice. I am not saying that these are good things, but the poverty causes of child labor are very, very different from the institutional causes of forced labor. When we lump these two together, we are not really serving either of these two problems as comprehensively as we need to."

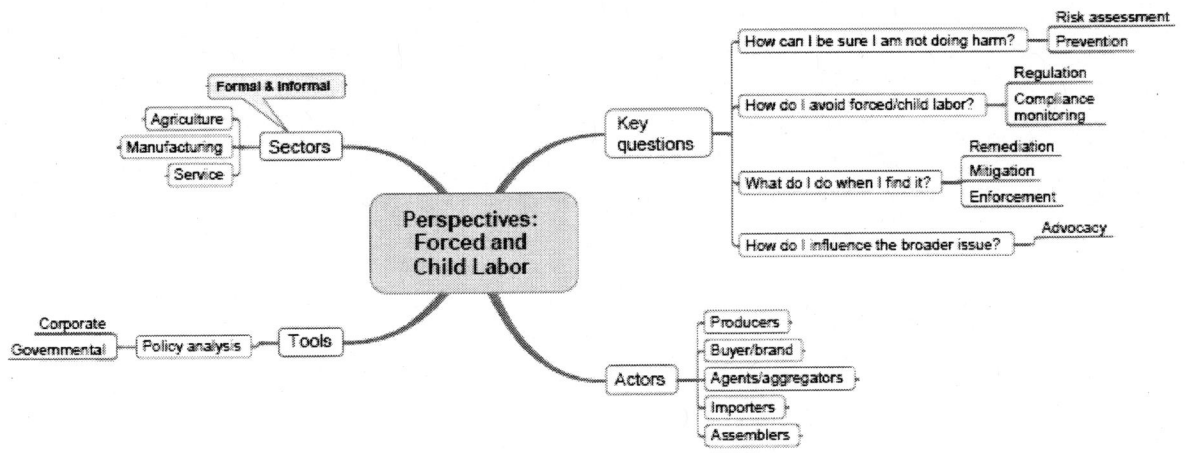

FIGURE 3-1 Perspectives: Forced and child labor.
Source: Susan Berkowitz, Chair, Planning Committee, Workshop Presentation, May 11, 2009

Second, it is important to consider migration and trafficking issues. Thea Lee, in her presentation, noted that with respect to forced labor it was important to consider how vulnerable migrant workers are to forced labor and trafficking. Her concern was that dealing with one particular forced labor situation (e.g., a brothel) could simply send the workers to another place without causing any positive effect.

Third, often one labor problem goes hand in hand with another, so it is important to consider the broader context of international labor standards, which include child and forced labor. Thea Lee emphasized this in her presentation, arguing that it is not possible to do a good job of evaluating child and forced labor or working to eliminate child and forced labor if you cannot also talk about freedom of association, the right to organize and bargain collectively, and protections against discrimination. Toni Dembski suggested that "when you are dealing with child and forced labor issues, you are also going to have contractors who are bending the rules in other places." Benjamin Smith from the ILO noted: "The inclusion of all of the fundamental principles and rights at work in multi-stakeholder initiatives and in efforts to clean up supply chains is really important as problems with the lack of freedom of association and collective bargaining or discrimination also contribute in a significant way to forced labor and child labor and vice versa. The interconnectedness of the fundamental principles and rights of work is reaffirmed in the 2008 Declaration on Social Justice for a Fair Globalization[1] and is something that is very important to keep in mind in looking at these systems." Problems in one labor area could be an indicator of possible problems in child or forced labor. Dan Viederman suggested integrating investigations for child and forced labor with the other investigations being done, learning how one indicator for one problem might lead to other indicators.

---

[1] Text of the Declaration available at: *http://www.ilo.org/wcmsp5/groups/public/---dgreports/---cabinet/documents/publication/wcms_099766.pdf* (accessed September 28, 2009).

Sectors

Another dimension of the context consists of sectors (or types of products and means of production) where child or forced labor might be found. Two concepts were discussed: (1) the differentiation among employment sectors and (2) the distinction between the formal and informal sectors or economies. A commonly applied set of standards is the United Nation's International Standard Industrial Classification of All Economic Activities (ISIC).[2] The classification breaks economic activities into 23 high-level categories that are further disaggregated into divisions and groups. For example, "agriculture, forestry, and fishing" is a high-level category that includes 3 divisions and 13 groups.

The committee, according to Dr. Berkowitz, "had a fairly elaborate breakdown of sectors, but we finally decided that for simplicity's sake, rather than try to find every last industry or every last farm activity we would just keep it at a more general level of looking at agriculture, manufacturing, and service sectors and both informal and formal activities within each of those sectors." According to the ILO, "The agricultural sector comprises activities in agriculture, hunting, forestry and fishing. The industry sector consists of mining and quarrying, manufacturing, construction and public utilities (electricity, gas and water). The services sector includes wholesale and retail trade, restaurants and hotels, transport, storage and communications, finance, insurance, real estate and business services, and community, social and personal services."[3]

Employment sectors can also be broken down into the formal and informal sectors. Formal-sector employment is often associated with salaried or wage-based labor which is typically captured in national labor statistics and subject to national laws and legislation. The informal sector, by contrast, includes those activities outside the formal sector, both legal and illegal, such as trade in stolen goods, prostitution, or wages from unreported work.[4] A related concern has to do with counterfeited goods. As one participant noted, there are many such goods produced in the world. Companies may be working to reduce child labor in their factories, while others are producing what appear to be the same goods in less protected labor situations. In any case, an important point is that the formal and informal sectors operate side by side in countries and the informal sector can constitute a large part of a country's economy, sometimes even most of it.[5] Much child and forced labor occurs in the informal sector.

---

[2] Revision 4 was adopted in 2008. See United Nations Statistics Division, Classifications Registry at: *http://unstats.un.org/unsd/cr/registry/isic-4.asp* (accessed September 28, 2009).

[3] ILO, *The End of Child Labour: Within reach,* pp. 7-8. "The agricultural sector comprises activities in agriculture, hunting, forestry and fishing. The industry sector consists of mining and quarrying, manufacturing, construction and public utilities (electricity, gas and water). The services sector includes wholesale and retail trade, restaurants and hotels, transport, storage and communications, finance, insurance, real estate and business services, and community, social and personal services."

[4] Paul E. Bangasser, The ILO and the informal sector: an institutional history. Geneva: ILO, 2000. World Bank, "Concept of Informal Sector" Available at: *http://lnweb90.worldbank.org/eca/eca.nsf/1f3aa35cab9dea4 f85256a77004e4ef4/2e4ede543787a0c085256a940073f4e4?OpenDocument* (accessed September 28, 2009).

[5] See for example: Friedrich Schneider, "Size and Measurement of the Informal Economy in 110 Countries Around the World." Paper presented at a Workshop of Australian National Tax Centre, ANU, Canberra, Australia, July 17, 2002. Available at: *http://rru.worldbank.org/Documents/PapersLinks/informal_economy.pdf* (accessed September 28, 2009).

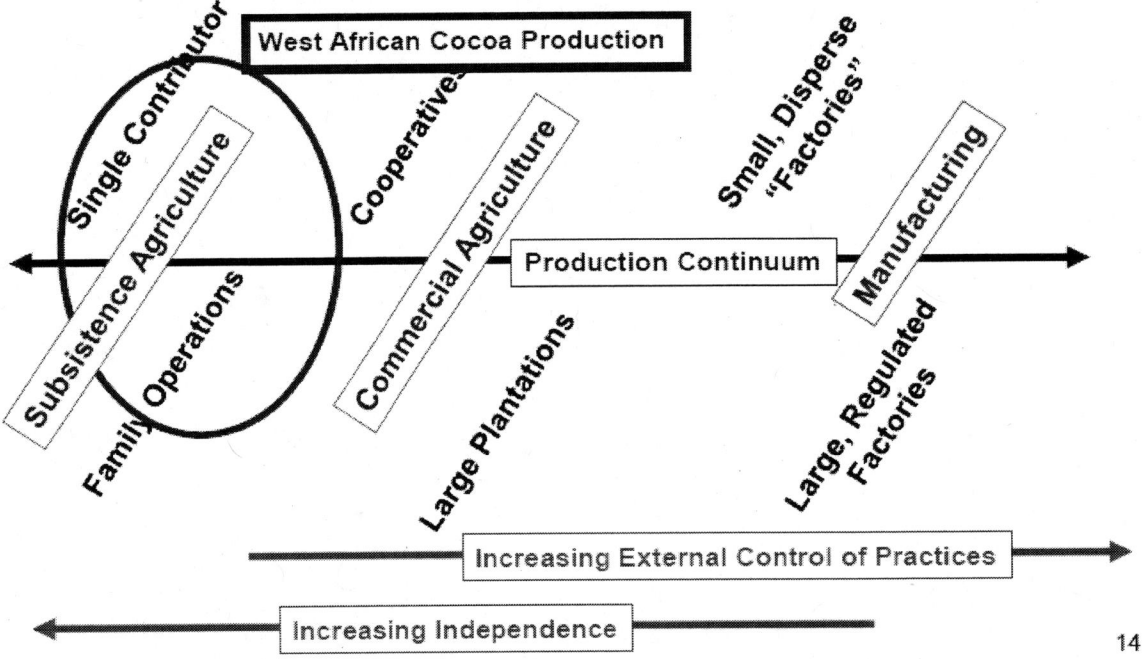

FIGURE 3-2 Defining the challenge and context in West African cocoa production.
Source: Jeffrey Morgan, Mars, Inc., Workshop Presentation, May 11, 2009.

Participants in the workshop debated how aggregated one should consider economic sectors. Jeff Morgan, in his presentation, argued that agriculture might be too broad and suggested one should disaggregate the sector, in particular to differentiate between subsistence and commercial farming. Mr. Morgan argued that there is a relationship between the employment sector and how the actors are organized, which is relevant for discussing the criteria. In Figure 3-2, he laid out a continuum for production from agriculture to manufacturing. He gave the example of banana plantations in Central America as contrasting significantly with cocoa farms in West Africa, due in part to their size and organization.

A second point of discussion was that different sectors can occur in the same geographic location and different labor practices can occur in the same sector but be organized differently. Kevin Bales gave the example of gold mining in Ghana, where you have tiny artisanal operations with the worst forms of child labor and forced labor within a quarter mile of a large, regulated international company's factory mine.[6] One is operating with medieval technology and one is operating with 21st-century technology and has careful safety controls and so forth. Mr. Bales noted that this provides an immediate alternative to the less attractive situation. It may be that

---

[6] See for instance: ILO-IPEC, *Prevention and Elimination of Child Labor in Mining in West Africa*, Project Document, Geneva, September 30, 2005. Gavin Hilson, "Challenges with Eradicating Child labour in the Artisanal Mining Sector: A case study of the Talensi-Nabdam District, Upper East Region of Ghana. Undated. Available at http://www.yorku.ca/cerlac/EI/papers/Hilson.pdf (accessed September 28, 2009).

workers can be moved into safer environments, with better labor protections, and that the workers need new skill training to be able to work in a better environment.

Actors

The focus of this report is on business practices,[7] but getting a product to market often involves a series of relationships among business entities that have a range of characteristics, including domestic and multinational enterprises or playing a range of roles as employers, producers, purchasers, vendors, or contractors. In confronting child or forced labor the focus sometimes seems to be on large multinational companies because they have the potential for significant reach and are connected to many other actors. By the same token, these companies may be somewhat removed from day-to-day operations of their various factories, vendors, and subcontractors. In this section we focus on workshop presentations that covered company supply chains and business engagement with external partners.

Supply chains

Supply chains represent the processes by which goods are produced from raw materials. Figure 3-3, for example, illustrates a relatively simple supply chain for soccer ball production. Jeff Morgan, from Mars Inc. presented as another example the cocoa supply chains originating in Cote d'Ivoire and Ghana (Figure 3-4).

Mr. Morgan commented that the cocoa sector is characterized by an extensive, complex, and disconnected supply chain. Unlike coffee, cocoa is primarily a small-holder crop in West Africa with over 1 million family farms producing cocoa and few farms belonging to co-operatives. Estimates are that in Côte d'Ivoire, less then 15 percent of the farmers would be members of a co-op and probably less than half of those are financially viable. In Ghana, the number of farmers and co-ops may be a bit larger but there is only one major co-op, Kuapa Kokoo.[8] Beyond the farm, the supply chain differs between Cote d'Ivoire and Ghana. The Ghanaian government sets the price and handles all of the transactions as they collect cocoa from the farm.

---

[7] An issue that is not explored is the degree of business responsibility or accountability for solving the problem of child or forced labor. This is an important issue but outside the scope of this report. A second issue not explored in this report but raised by a participant during the discussion had to do with how best to encourage businesses to take action. The participant saw a tension between using the media to expose problems in companies' supply chains, when often these high-profile companies were the ones taking the best steps among companies to deal with the problem. Additionally, the participant wondered whether the threat of criticism would lead companies to want to keep problems quiet so as not to invite The criticism. A few of the presenters thought that companies would be willing to share best practices and certainly would be willing to work together to improve the labor situation. As Anna Walker of Levi's noted, "If we have a critical mass in our industry that is following the same standards and applying the same conditions and then having the same dialogue, it's just going to make life easier for us." Another participant noted that companies' codes of conduct and reports were widely available on the Internet.

[8] A cocoa producer cooperative founded in Ghana in 1993. See Kuapa Kokoo at *http://kuapakokoogh.com/kuapa/* (accessed September 28, 2009).See also Transfair USA, "Fair Trade Cocoa Co-op." available at *http://www.transfairusa.org/content/certification/producer.php?floid=1475* (accessed September 28, 2009).

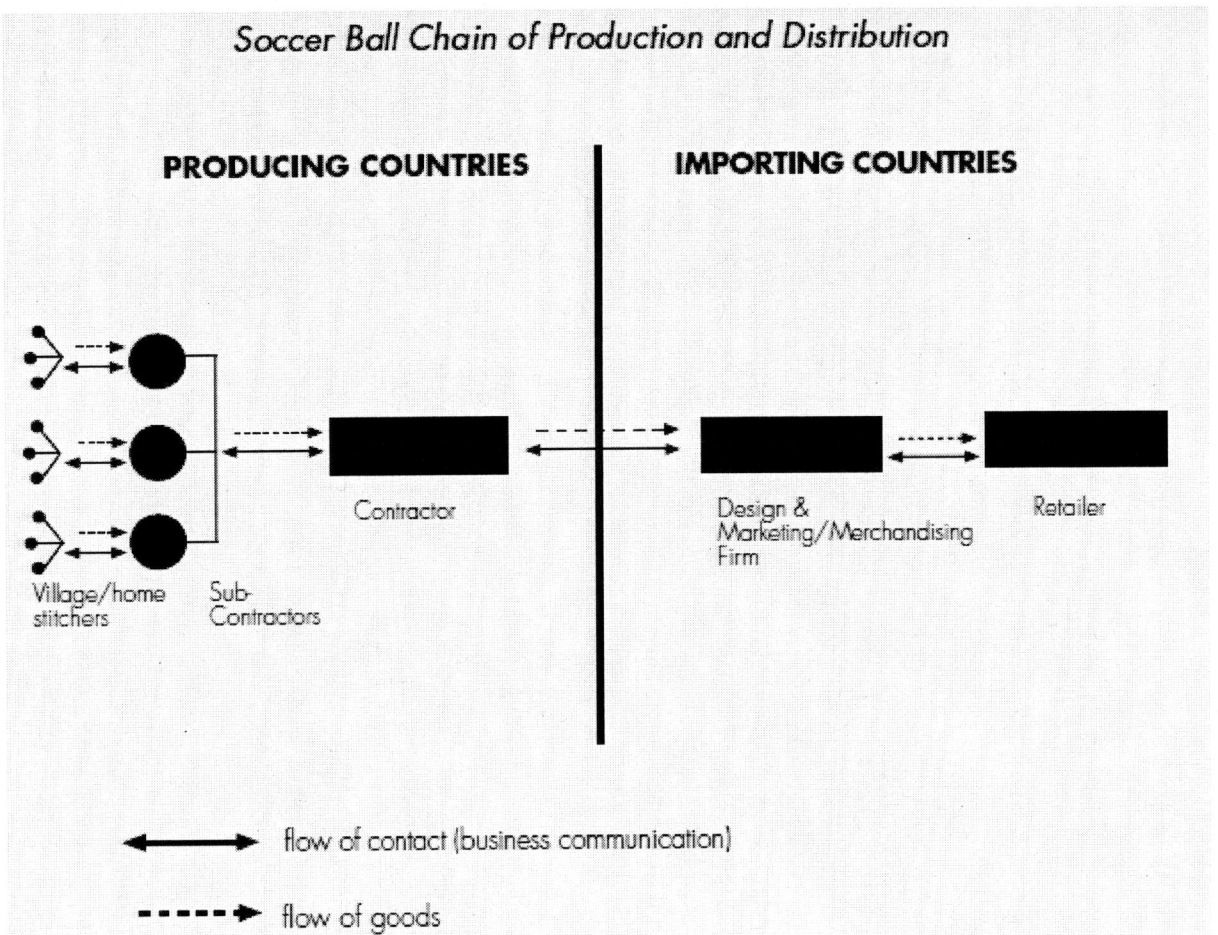

FIGURE 3-3 Soccer ball production and distribution.
Source: U.S. Department of Labor, *By the Sweat and Toil of Children (Volume IV): Consumer Labels and Child Labor*, 1997, p. 99.

Mr. Morgan drew particular attention to the area below the horizontal line in Figure 3-4 that focuses on the global market and the manufacturer, noting that companies like Mars do not take possession of cocoa until after the good has been exported. Along the supply chain the cocoa will change ownership at least six times and often many times more. The challenge facing Mars has been how to affect behavior on a family farm that is a great distance removed in the supply chain.

Ms. Hauchère also gave an example of a soya supply chain in Brazil (Figure 3-5). At the request of the government Reporter Brasil, an NGO, with the support of the ILO, identified the production and supply chains of an estate on the "dirty list" of employers involved in forced labor. Two studies were done, in 2004 and in 2007, mapping estates appearing on the dirty list and the companies working with them. Ms. Hauchere noted that she chose the soya example not because it is a supply chain with substantial forced labor but because it is relatively simple. Most forced labor cases in Brazil are found in the meat processing supply chain, which is too complex to be represented like this.

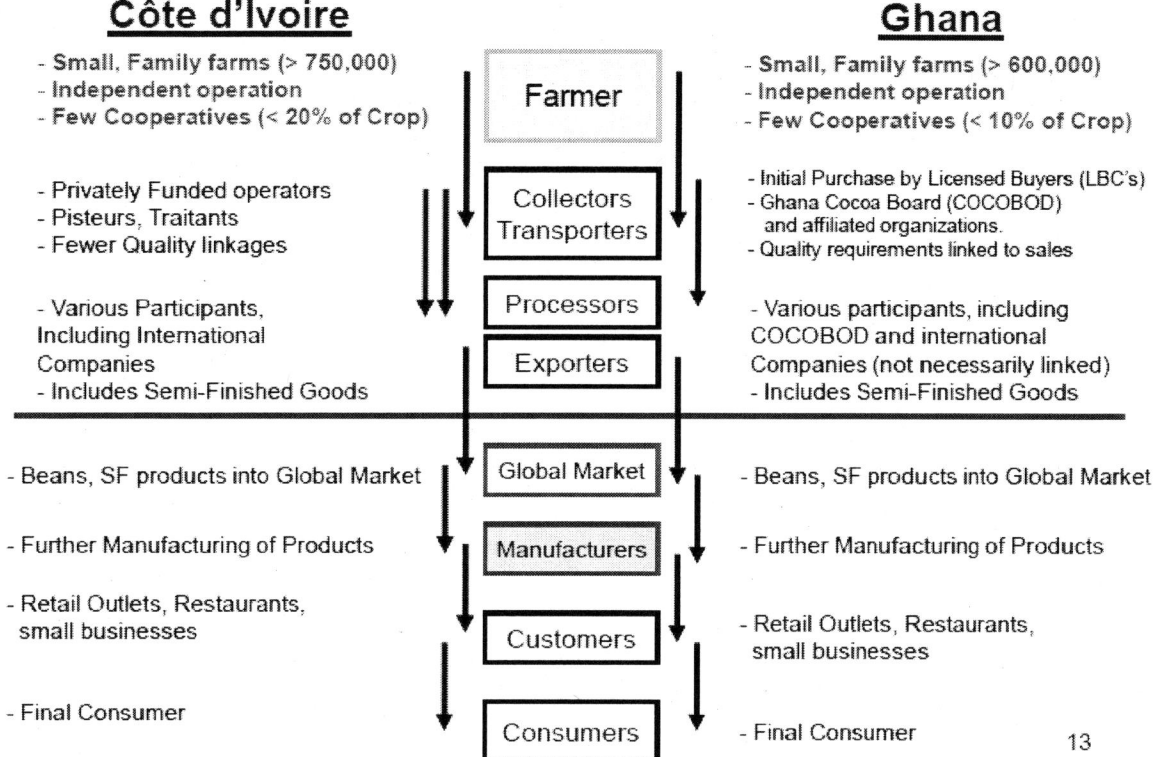

FIGURE 3-4 Cocoa supply chains in Cote d'Ivoire and Ghana.
Source: Jeffrey Morgan, Mars Inc., Workshop Presentation, May 11, 2009.

Partnerships

Presenters and audience members noted that there are any number of partnerships that can be formed between business and other entities, including governments, other companies, NGOs, trade unions, workers, and local communities in order to address child and forced labor. Presenters such as Bill Guyton, with the World Cocoa Foundation, noted that many groups were involved in the effort. To look just at the industry or one group to fix the problem is not going to work. You need to bring all of the players to the table together, and that is where agencies like U.S. State Department and DOL can help by being the convener, inviting governments, civil society members, and industry to sit down together. We have seen that happen on several occasions. That is the right path forward.

Presenters and audience members mentioned the important role of government. Jeff Morgan commented that in the case of cocoa, in both Ghana and Cote d'Ivoire the government is highly engaged in the sector. They grant cocoa purchasing licenses and the sector has historically been a source of much tax revenue. The Cote d'Ivoire government is quite strict in allowing actors to go into the country and work in communities. Mr. Morgan noted that one cannot do that without the approval of the government and trying to find out who gives that approval can be challenging, given the less than stable current governance situation in the country. Ms. Roggensack noted in her presentation (detailed in the next chapter) that countries need to adhere to the international labor standards. Mark Neuman reminded the audience that

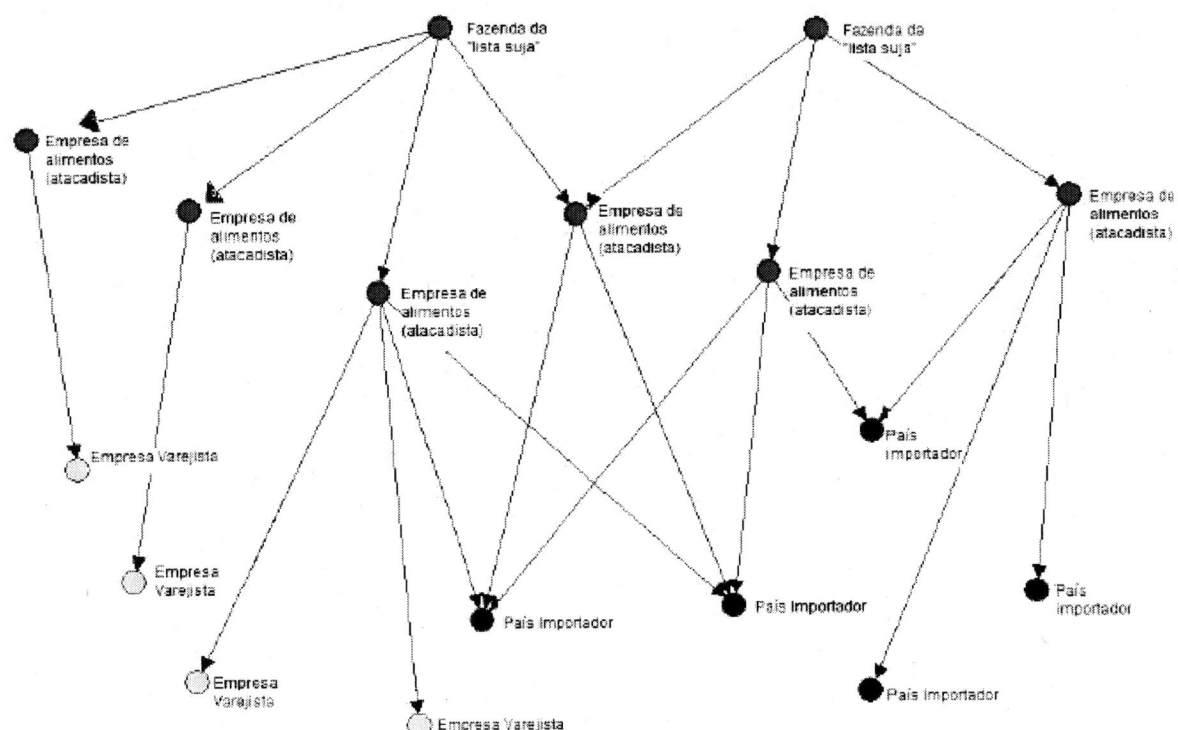

FIGURE 3-5 Soya supply chain in Brazil. The red dots are estates on the dirty list. The blue dots are wholesalers buying from those estates but not selling directly to customers. At the next level are two types of companies: the yellow ones selling directly on the national market; the black are exporting companies.
Source: Aurelie Hauchère, International Labour Organisation, Workshop Presentation, May 12, 2009. See also Patricia Trinidade Maranhão Costa, Fighting Forced Labour: The Example of Brazil, Geneva: ILO: Special Action Programme to Combat Forced Labour, 2009, p. 94.

it is not just the country where child or forced labor may occur that should act, but also the trading nations that could leverage their influence to make a difference.

Mark Neuman began his presentation by noting that child labor is a well-known problem in Uzbekistan.[9] Even though the Uzbek government signed ILO convention No. 182 in 2008, it has not been taking steps to implement the convention, according to Mr. Neuman. Mr. Neuman suggested that the U.S. government could do more to influence the Uzbek government, however,

---

[9] See for instance, Environmental Justice Foundation. *The Children Behind Our Cotton*. London, UK: Environmental Justice Foundation, 2007; U.S. Department of Labor, *The Department of Labor's 2007 Findings on the Worst Forms of Child Labor*. Washington, DC: U.S. Department of Labor, 2008; Uzbek Human Rights Activists and Journalists, Forced Child Labor in Uzbekistan's 2007 Cotton Harvest: Survey Results. 2008. Available at: *http://www.laborrights.org/files/Survey%20report%20on%20child%20labour%20in%20Uzbekistan%20April%202008.pdf* (accessed September 28, 2009).

he was critical of U.S. legislation and the generalized system of preferences. Neuman noted that 19 USC §1307 of the U.S. Code prohibited the importation of goods made by forced labor, but an exception clause[10] could be seen as an escape clause and he suggested that this loophole needs to be closed. He also noted that 19 USC §2462 (the Generalized System of Preferences program) makes countries ineligible for designation as beneficiaries of the program if they fail to implement their commitments to eliminate the worst forms of child labor. Neuman noted that the United States imported $3 million in mined and agricultural products under the GSP program from Uzbekistan in 2008. Neuman further contended that there was a lack of coordination between various U.S. agencies charged with focusing on labor issues, including: DOL, U.S. Department of Agriculture, U.S. Department of State, and U.S. Department of Homeland Security (Customs and Border Protection and Immigration and Customs Enforcement). The U.S. government could act in a more coordinated way, and with its various laws and enforcement capabilities, the government could affect child labor in Uzbekistan.[11]

Tools

A third dimension of the context pertains to tools that could be employed to combat child and forced labor. There are a number of types of business practice.

The ILO provides one typology, stating "The range of actions on child labour taken by employers and their organizations to date can be broken down into the following categories:

- general awareness-raising and policy development initiatives;
- prevention of child labour in specific sectors;
- direct support for initiatives aimed at the removal and rehabilitation of child workers;
- certification schemes for specific goods, and
- corporate and industry codes of conduct."[12]

The International Finance Corporation identifies several steps companies can take:

- addressing the issue in the workplace:
    - awareness raising;
    - policy development (including identifying and complying with relevant laws; focusing on positive incentives);
    - implementation (for example, establishing procedures and practices, providing training, building accountability);
    - monitoring;, and
    - forging partnerships with other companies and organizations.

---

[10] In no case shall such provisions be applicable to goods, wares, articles, or merchandise so mined, produced, or manufactured which are not mined, produced, or manufactured in such quantities in the United States as to meet the consumptive demands of the United States.

[11] Uzbekistan had submitted an instrument of ratification regarding Convention No. 138 in July 2008, but it was not registered until June 3, 2009. (See: *http://webfusion.ilo.org/public/db/standards/normes/appl/index.cfm?lang=EN*) (accessed September 28, 2009).

[12] International Program on the Elimination of Child Labour, Action Against Child Labour: Lessons and Strategic Priorities for the Future, A Synthesis Report. Geneva: ILO: 1997, p. 228.

- responding when child labor is detected; and
- managing risks in companies' supply chains, via
    - selecting quality suppliers;
    - prohibition clauses in contractual agreements;
    - establishing subcontracting safeguards;
    - labeling and certification;
    - consolidating production centers;
    - providing supplier training and incentives;
    - monitoring, compliance and corrective action; and
    - scaling up efforts with suppliers.[13]

More recently, ILO developed 10 principles for business leaders to combat forced labor and trafficking:[14]

1. Have a clear and transparent company policy, setting out the measures taken to prevent forced labor and trafficking. Clarify that the policy applies to all enterprises involved in a company's product and supply chains.
2. Train auditors, human resource, and compliance officers to identify forced labor in practice and seek appropriate remedies.
3. Provide regular information to shareholders and potential investors, attracting them to products and services where there is a clear and sustainable commitment to ethical business practice, including prevention of forced labor.
4. Promote agreements and codes of conduct by the industrial sector (as in agriculture, construction, and textiles), identifying the areas where there is risk of forced labor, and take appropriate remedial measures.
5. Treat migrant workers fairly. Monitor carefully the agencies that provide contract labor, especially across borders, blacklisting those known to have used abusive practices and forced labor.
6. Ensure that all workers have written contracts in language they can easily understand, specifying their rights with regard to payment of wages, overtime, retention of identity documents, and other issues related to preventing forced labor.
7. Encourage national and international events among business actors, identifying potential problem areas, and sharing good practice.
8. Contribute to programs and projects to assist, through vocational training and other appropriate measures, the victims of forced labor and trafficking.
9. Build bridges between governments, workers, law enforcement agencies, and labor inspectorates, promoting cooperation in action against forced labor and trafficking.
10. Find innovative means to reward good practice, in conjunction with the media.

There are commonalities across each of these examples of classificatory schemes, including:

- Businesses have multiple avenues to affect child or forced labor.

---

[13] IFC, Addressing Child Labor in the Workplace and Supply Chain. Good Practice Note. Number 1. June 2002.
[14] ILO, Strengthening Employers' Activities against Forced Labour, 2008, p. 3.

- Businesses can use both positive and negative incentives to tackle these problems, as noted by a few audience members during the workshop.
- Businesses can operate both by themselves and in partnership with a variety of other partners.
- There are a number of different points where businesses can apply leverage.

Key Questions

Having examined many dimensions of the context, Dr. Berkowitz also presented several key questions the planning committee identified as important for assessing business practice:

- *How can I be sure I am not doing harm?* Dr. Berkowitz noted that the planning committee identified risk assessment and prevention activities as additional components related to the context that businesses should engage in as part of any business plan targeting child or forced labor.
- *How do I avoid the use of forced and child labor?* Dr. Berkowitz noted that compliance monitoring and regulation are sets of activities that would be responsive to this question.
- *What do I do when I find child or forced labor?* Dr. Berkowitz indicated that remediation, mitigation and enforcement are key business practices that would come into play in response to findings from this question.
- *How do I influence the broader issue?* Businesses can address their specific situations in the instances where they find problems, but they may also combat child and forced labor through broader action.

Additional Comments

Several participants felt that a contextual element missing from the discussion concerned the overall environment of the country where child or forced labor occurs, including such elements as the rule of law, relevant legislation, governance, taxation, and economic development. Examples of comments follow:

- Kevin Bales suggested that the most powerful predictor of whether or not there are forced labor violations is the quality and content of the rule of law.
- One audience member asked whether the country even recognized the ILO Conventions. Has it signed them? Is it actively trying to push for adherence to those Conventions? Is there a functioning Department of Labor? Are there activities in the field? Are there deputy ministers who have staff that have ever visited the field?
- Thea Lee asked whether there is enough attention being paid to the underlying economic forces, the policy issues, the labor law, implementation, the labor law reform that might be needed in order to get compliance. Is the legal structure adequate to do the job? Are the laws enforced? Does the government have adequate resources for enforcement and is the government using those effectively?
- Another audience member added, I am also wondering if we should not have something about the impact on the wider population and the wider economy, or the wider society

when an economic boycott is used to intervene in a forced labor or child labor situation. For example, Uzbekistan has serious problems with forced labor in cotton production, but if there were a boycott what would happen to those families who obviously are not responsible for this but in fact would be very significantly, very powerfully, detrimentally affected by an economic boycott of the primary export product of that country? While the government is certainly culpable, the average citizen is possibly not and may not be able to agree to the idea that they should be asked to sacrifice significantly because of our concerns about what happens in producing their product. That is an illustration of what I mean by the broader impact.

- One participant noted that it was critical to get at root causes. This individual described a case where a girl was found working in a factory—a case of child labor. The company offered to pay for the girl's schooling. For whatever reasons the family had—the participant speculated it was underlying poverty—the girl continued to work.
- Another noted that in Côte d'Ivoire one of the big problems is that basically farmers get to retain very little money for what they produce. They are taxed for more than two-thirds of what they produce, so they have to make children work, otherwise they cannot make ends meet. We may very well build the schools or organize the community, but the problem is with the government and with all of the corruption surrounding the cocoa industry.
- Jeff Morgan noted that many cocoa farmers do not use modern farming techniques and, as a result, yields are lower than they could be. With the combined pressures of taxation and little access to loans, farmers do not have the resources to hire labor. Farm labor is thus more likely to fall on the family.

4
ILLUSTRATIVE BUSINESS PRACTICES

Introduction

Dealing with child or forced labor is an imperative for businesses for a variety of reasons. The International Organization of Employers (IOE) notes the obvious ethical or moral considerations and argues that businesses have a responsibility to mitigate child or forced labor.

> Where a company becomes engaged in the issue of child labour, its priority should be to work towards the complete and unconditional elimination of the WFCL [worst forms of child labor]. This should also include measures to address the needs of the affected children's families. Once the WFCL have been completely addressed in a particular context, business can, where it is within its means and power to do so, explore efforts to promote beneficial conditions in the context of the other forms of child labour as a next step. Again, these measures should address the needs of the children's families.[1]

There are also economic reasons for business to address the worst forms of child labor because such labor may have a negative impact on companies' bottom lines. Child laborers have fewer opportunities to go to school to become more skilled adult workers and child labor can undermine economic growth and depress the consumer market. Such labor may also affect the public's perception of the company. Businesses are in a unique position to become involved and can have a direct impact on child labor.[2]

The degree of responsibility businesses should take and the scope of remedies are a matter of opinion. In this section we present examples of business practices and practices followed by NGOs with some degree of business engagement, gleaned from presentations at the workshop. (Additional examples of business practices can be found in Appendix F.)

The Cocoa Sector

Three presentations—by Jeff Morgan of Mars Inc., Bill Guyton of the World Cocoa Foundation, and Meg Roggensack of Free the Slaves—focused on cocoa as an example of a sector in which child labor is known to occur and where industry has taken steps alone and in various partnerships.

Bill Guyton began his presentation by noting that there are 1.5 to 2 million small-scale family farms in West and Central Africa, stretching from Cote d'Ivoire to Cameroon. About 10 million people live on cocoa farms and the cocoa supplies about 50 percent of the household income, particularly in Côte d'Ivoire. While referred to as

---

[1] International Organization of Employers. Challenges in Addressing Child Labour: IOE, May 2005.
[2] Ibid.

cocoa farms, generally these family farms grow multiple crops. Mr. Guyton noted that the World Cocoa Foundation[3] has many member organizations including large branded companies who participate in the programs, processors, traders, port authorities, and many smaller chocolate companies from such areas as North America, Europe, and Southeast Asia. He also noted that the World Cocoa Foundation partnered with many more organizations:

- Farmers and farmer organizations
- Governments in cocoa producing countries
- Government and development agencies
  - CGIAR Centers (IITA and ICRAF)
  - GTZ (Germany)
  - Danida (Denmark)
  - U.S. Agency For International Development (USAID)
  - U.S. Department of Agriculture (USDA)
- NGOs (Winrock, ACDI/VOCA, IFESH, Technoserve, Socodevi, Making Cents)
- Researchers and universities

Mr. Morgan began his presentation by offering background on child labor issues in West African cocoa production. He noted that the industry began to be pressed on this issue in September of 2000 and more so in the spring of 2001 when there were public reports of children in very dire straights in the cocoa sector. He highlighted such information as the September 2000 documentary on Channel 4 in the United Kingdom and the spring 2001 series of articles in the *Philadelphia Inquirer* that featured children who had been trafficked and were forced to work in absolutely unacceptable conditions. Mr. Morgan commented that in his view "part of the concern that we in industry had was that although these were anecdotal accounts, which we all have to deal with, I think some of the comments on their prevalence, we felt were exaggerated."

Ms. Roggensack further framed the challenge in the cocoa sector. Research and surveys show that the predominant phenomenon is the engagement of children of families in hazardous work activities on family-owned cocoa farms: carrying heavy loads, spraying hazardous chemicals, and using machetes to open cocoa pods. She noted that this is exacerbated by poverty in this area, and a lack of awareness. "We are talking about a head of household, a parent's lack of awareness that they are putting their child at risk of harm but over the longer term that they are harming their welfare, their ability to grow and develop." She added the factor of tradition, the fact that many of the individuals, practically all of them, grew up in the same cultural milieu. This is how it has always been done and there is the notion that there is some value in learning how to do this because this is the child's future.

What the studies do find, she suggested, is that simply removing children from cocoa farming activities is not necessarily the entire answer. It may lead to displacement, to other activities such as domestic work, or other forms of work that could be equally damaging if not as hazardous to welfare. If not twinned with a quality educational alternative, removing the children is not going to result in the change that you want.

---

[3] See "About the World Cocoa Foundation" at http://www.worldcocoafoundation.org/about/default.asp.

Mr. Morgan noted that the industry did react.[4] The U.S. industry collaborated with the offices of Senator Harkin, Representative Engel, and Senator Kohl during July 2001 to address the issue. Comparing the response in the United Kingdom to that in the United States, Mr. Morgan remarked that in the UK it was more of a consumer and consumer agency issue, whereas in the United States, it became a legislative issue led by members of Congress who were starting to call for a labeling provision. The industry's view in Mr. Morgan's perspective was that such action would have been damaging to both the industry and the cocoa farms. Industry signed on to the Harkin-Engel Protocol.[5] Since that time, the cocoa industry has been pursuing a consolidated effort to address child and adult labor issues in the cocoa sectors of Ghana and Côte d'Ivoire. According to Mr. Morgan, the only remaining step is the implementation of standards of public certification in the cocoa sector.

Mr. Morgan said that the Protocol required that "industry in partnership with other major stakeholders will develop and implement credible ... standards of public certification ... that cocoa beans and their derivative products are grown ...without any of the worst forms of child labor." This was to be accomplished by July 1, 2005, but industry did not meet this deadline. New milestones were proposed and agreed to. The industry was to implement a certification system covering 50 percent of the cocoa-growing areas of Ghana and Côte d'Ivoire by July 2008, with industry dedicating more than $5 million annually to this effort. The milestone for 2010 is independent verification of sector-wide reports from Côte d'Ivoire and Ghana (see Figure 4-1).

---

[4] There are currently a number of actors involved in this issue: Cadbury, Ferrero, The Hershey Company, Kraft Foods, Mars Incorporated, Nestlé, ADM Cocoa, Barry Callebaut, Cargill, Association of the Chocolate, Biscuit, and Confectionery Industries of the EU (CAOBISCO), Confectionery Manufacturers of Australasia (CMA), Confectionery Manufacturers Association of Canada (CMAC), European Cocoa Association (ECA), Federation of Cocoa Commerce (FCC), National Confectioners Association of the US (NCA), and the World Cocoa Foundation (WCF). There have been a few critiques of industry's response to child labor and cocoa. See for instance: the ILRF, "The Cocoa Protocol: Success or Failure? June 30, 2008, available at http://www.laborrights.org/sites/default/files/publications-and-resources/Cocoa%20Protocol%20Success%20or%20Failure%20June%202008.pdf.

[5] For the text of the protocol, see: *http://www.cocoainitiative.org/images/stories/pdf/harkin%20engel%20protocol.pdf.* See also "Joint Statement from U.S. Senator Tom Harkin, Representative Eliot Engel and the Chocolate/Cocoa Industry on Efforts to Address the Worst Forms of Child Labor in Cocoa Growing Protocol Work Continues." July 1, 2005. Available at: http://harkin.senate.gov/pr/p.cfm?i=240245.

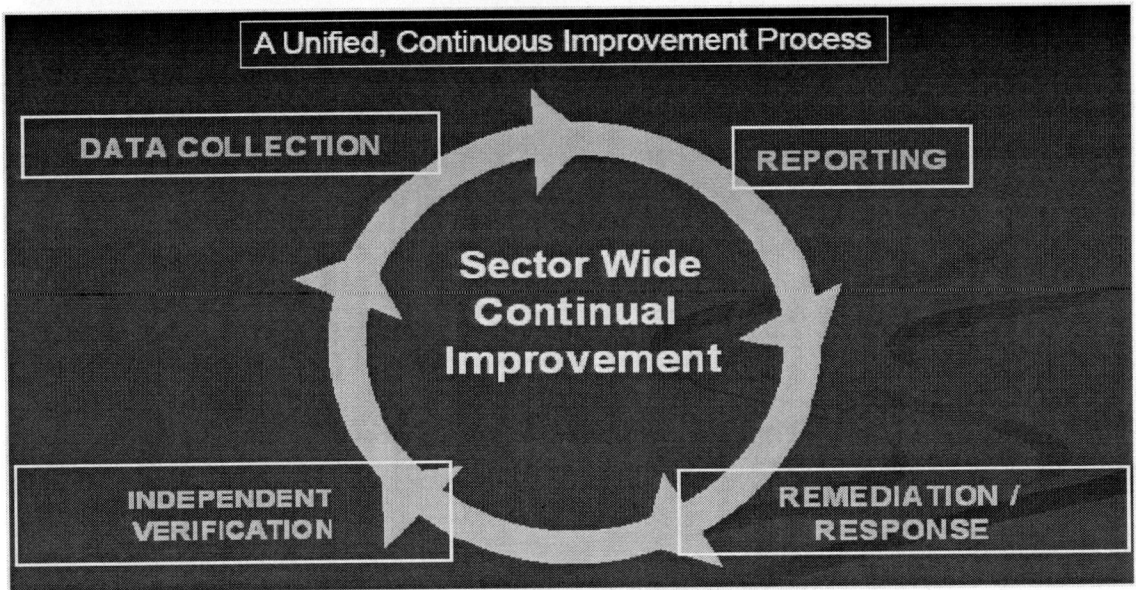

Figure 4-1 Cocoa sector country certification model implemented in conjunction with national government programs.
Source: Jeffrey Morgan, Mars, Inc. Workshop Presentation, May 11, 2009.

The Cocoa Industry Certification Program

Jeff Morgan of Mars Inc. noted that the cocoa industry certification program is a complex model that industry was trying to implement across the cocoa- producing sectors of two countries: Ghana and Côte d'Ivoire. The program comprises data collection, public reporting, remediation and response, and an independent third- party verification.

Industry has sought credible data on child and adult labor practices in the regions. Concerning remediation, both data collection and remediation require significant financing. Noting that there were likely over 1 million independent farms in Ghana and Côte d'Ivoire, Mr. Morgan suggested that one could spend enormous amounts of money trying to evaluate what is happening on each farm and then have little left over for remediation. We should try to balance the findings and the data that are needed and put the majority of the resources in remediation.

Concerning verification, Mr. Morgan noted that stakeholders are skeptical that information brought forward by the industry and the governments in this area is untainted. He said that industry, under the Harkin-Engel Protocol, has been working on implementing an independent third-party verification process.

Mr. Morgan commented on the impact of the practices to date:

- There have been a series of surveys completed both in Ghana and Côte d'Ivoire. In May 2009, he noted, members of stakeholders would meet in Accra, Ghana, to determine the best step way forward regarding the quality of data and how much

data collection is really needed as the industry goes forward.
- Reports detailing the incidence of child labor and adult labor have been posted on Web sites with the support of the governments of Ghana and Côte d'Ivoire. Mr. Morgan commented: "I think this is a real breakthrough. When we began the process, many in the governments were hesitant to agree that there was a child labor issue and now we have reports on their Web sites that detail the extent of the issue."
- A portfolio of activities is being undertaken by industry vis-à-vis remediation to address this issue.
- There was a third-party verification in December 2008. There have been other meetings, and dissemination of those findings in West Africa and the European Union and in the United States. Tulane University conducted an oversight evaluation. According to Mr. Morgan: "To the best of our understanding of their results, they have had two sets of results publicized and they are finding very similar results to the surveys that have been published by the countries."

Mr. Guyton spoke about the World Cocoa Foundation's programs. He began by addressing some of the most important characteristics of those programs over the last several years:

- Effective public-private-partnerships;
- Empowerment of smallholder farmers;
- Host country engagement and capacity development;
- Economic, social and environmental components;
- Continuity of approach and resources;
- Tangible results orientation;
- Scalability; and
- Pooling of resources and expertise.

The Sustainable Tree Crops Program

Mr. Guyton identified several goals for the WCF's programs, including the one most relevant to this workshop: National and international labor standards are implemented with no worst forms of child labor. He detailed two major programs that the World Cocoa Foundation has in West Africa: (1) the Sustainable Tree Crops Program that covers five West African countries (Cote d'Ivoire, Ghana, Cameroon, Nigeria, and Liberia) and (2) a public-private partnership called ECHOES that is specifically looking at youth empowerment, youth livelihood and education in the countries of Ghana and Côte d'Ivoire. Details on this program were presented by Vicki Walker of Winrock, and are summarized later in this chapter.

The Sustainable Tree Crops program intends to improve farm incomes by up to 50 percent, encourage safe farming practices, increase awareness and understanding of child labor issues, and ensure that labor standards are incorporated in all WCF programs. The program components include: farmer field schools, the primary method of reaching out to farmer groups; farmer organizational support through which WCF is helping to develop farmer co-ops to strengthen them in all five countries; and approaches to improved marketing practices as well as policy and research.

Farmer field schools are conducted in the farmer fields and they look at how to improve the productivity of the farms, farm safety, child labor awareness, HIV/AIDS and malaria prevention, marketing skills, gender-specific training and diversification.

With respect to metrics of success, Mr. Guyton noted that the program's goal is to reach 150,000 farmers. To date, the WCF has reached about 76,000 in the 5 countries. The WCF has seen through surveys that incomes of farmers who graduate from the farmer field schools increase 22 to 55 percent, depending upon the country. The WCF has seen (by surveys in Ghana) that for every 1,000 farmers trained, 210 children have been removed from hazardous work conditions.

Mr. Guyton included a photograph from a farmer field school in Cameroon. The farmers actually put together a cropping calendar of what they do at different months of the year and then they looked at who in the family actually carries out certain activities, whether it is the father, the mother, or the children. This is the first step in understanding child labor, and which sorts of activities are appropriate and not appropriate for children to be undertaking. This is matched with the school year to explain how important it is for the children to be in school rather than helping their parents in the fields during certain times of the year.

The International Cocoa Initiative

Ms. Roggensack described the experiences of the International Cocoa Initiative (ICI), which is one of the outcomes of the Harkin-Engel Protocol.[6] ICI is funded by industry and core funding is guaranteed. It is an independent, nongovernmental organization headquartered in Geneva, Switzerland. ICI is a multistakeholder activity that engages all parts of the production chain, the processors, the brands, and national governments. It is focused on the cocoa sector and in particular on the worst forms of child labor and forced labor.

The most important aspect of it is the role of community engagement. The first thing that ICI did was to go into communities on a pilot basis and talk to a wide range of individuals, including local leaders, women, and children, to listen to their perspectives and to sensitize them to the fact that what they were doing with regard to child labor was not appropriate. From that understanding ICI began a communication about the steps communities would like to take, what would be appropriate to their circumstances, and how their needs might be best met. ICI replicated this experience, widening it out to a larger category of communities, to create community action plans that include implementation and monitoring components as well. In virtually every community action plan one of the first things is the request for quality educational alternatives: teachers, rehabilitation, schools, and vocational training, something that Ms. Roggensack felt was an integral part of this model.

Ms. Roggensack noted that national legal norms and company practices were important to the work of ICI. When the ICI started its work, the relevant countries had ratified the pertinent ILO Conventions, but they were in denial about the challenge they were facing and at virtually every level ignorant about the pertinent standards and laws.

---

[6] See International Cocoa Initiative, "Home" available at *http://www.cocoainitiative.org/*.

There were no national plans. There was no hazard list and there was no active engagement by the authorities in Ghana or Côte d'Ivoire. One of the first things that the ICI accomplished was awareness raising aimed at the governments, business, and communities. The ICI then worked to provide training and technical support to the governments, for example, by assisting in the development of national plans or, with the ILO and foreign donors, for a list of hazardous work.

Ms. Roggensack noted that while companies involved in cocoa production tended to have codes of conduct that included some focus on child or forced labor, there were barriers to transmitting the codes all the way down the supply chain. The result was less coordination. The ICI provided a platform for businesses to come together, use their collective bargaining power and political power to make these labor issues a priority at the government level, allocate resources, and engage the cocoa boards and the licensed buying cooperatives.

A second challenge noted by Ms. Roggensack concerned collection of data on child labor. When the ICI began, the governments were not collecting quality data and a lot of misinformation was in the media. While businesses had undertaken internal audits, these efforts were criticized as neither independent nor particularly credible given some of the heated concerns around this issue. The ICI again played an important role in providing a platform for conducting initial pilot surveys that could be replicated. These surveys were valuable in developing a better sense of how these communities work.

A third area of focus for the ICI is remediation. Within the cocoa communities there was low enrollment in schools, few teachers, and little government contribution to teacher's salaries. Companies responded with such efforts as farmer field school training, but these efforts were not necessarily intended to be a coordinated response to a challenge. What ICI did through its engagement with the communities was to articulate a priority for education and in the process start to give voice to the members of the community. Additionally, the ICI helped the governments implement education policy changes.

A variety of reports have been released and disseminated on the Internet: the ICI's original pilot work, quarterly and annual reports, the certification survey results that the governments have published, the verification board's work, and the Tulane oversight studies.[7] Businesses have also published their own reports on what they are doing and in some cases what their foundations are doing; they also have their own individual outreach to their stakeholders.

Target

The case of the Target Corporation, a large-volume importer of thousands of items, contrasts with the case of cocoa, which is focused on a deep supply chain for a specific commodity. Toni Dembski, senior counsel with Target, presented Target's social responsibility program, "Global Compliance," part of which covers vendors and is connected to Target's sourcing and quality function.

Target operates 1,600 stores in the United States and will sell annually over 250,000 different items. Target is the number-two importer in the United States by

---

[7] Payson Center for International Development, Tulane University. Available at: *http://www.childlabor-payson.org/default.html*.

volume, importing $10 billion worth of merchandise across 70 of the 100 chapters in the U.S. tariff book. At any given time in Target's stores there are 60,000 to 80,000 items. Many of those items are trend items and will only be in the store for six weeks or so.

According to Ms. Dembski, Target has a world-class social responsibility program, called "Global Compliance," which is 10 years old.[8] Target's goal in developing its program was to be proactive in defining the program and its objectives. The program is sensitive to legal and labor issues as well as the social responsibility expectations of its customers.

Global Compliance covers more than child labor or forced labor though those were two of the initial problems. Given Target's breadth, its global compliance program reach extended significantly beyond those two issues, to include such areas as a safe and healthy workplace, nondiscriminatory practices, fair pay and wages, and compliance with several other laws of the United States.

Ms. Dembski noted that it is difficult for companies like Target that deal with thousands of vendors to probe far into the supply chain. Target's approach is to raise awareness both internally and with its vendors. Ms. Dembski believes that Target's position can give it leverage over its vendors: "If you want to work with Target, the number-two importer in the United States, you better understand what you are doing. You better understand our standards and actually apply those standards in the way you conduct your own business." Awareness raising is done through education and information is disseminated in a variety of formats, such as through Web sites, program materials, and training both online and in person.

Target trains internal teams to work with foreign vendors and conducts audits of factories all over the world. Whenever a vendor wants to do business with Target, it has to give Target access to all of its owned and nonowned facilities anywhere goods will be produced for Target. Those facilities initially answer questions about the working conditions and then they are measured on a series of different kinds of audits. For example, lawyers might visit the facility and look at particular issues, while other auditors might focus on social responsibility. Audits involve questionnaires and interviews with employees and managers. She noted that because of the large number of factories involved, Target randomly selects factories to be visited each year. Over a five-year period all factories are visited. If a vendor disregards Target's standards repeatedly, it will be eliminated from their pool of vendors.[9]

Levi Strauss And Company

Anna Walker, manager of government affairs and public policy, presented business practices at Levi Strauss. Levi's was founded in 1853 by Bavarian immigrant Levi Strauss who had a dry goods store and was servicing the "49'ers" mining gold in the Sierra Nevada and brought us the iconic jean. The company has grown to 11,000

---

[8] For a broader description of the program, please see Target Corporate Social Responsibility Report 2008 available at: *http://www.socialfunds.com/shared/reports/1229103541_Target_CR_Report_2008.pdf*.
[9] One participant from an NGO who had worked with Target noted that Target had strong incentives in place. Over time factories that were not in compliance were dropped from working with Target. Another participant asked about prison labor and Ms. Dembski noted that in the past Target had found that it was working with prison factories and had eliminated them from its factory base.

employees around the world, 1,500 of whom are at its headquarters in San Francisco. The company had revenues in 2008 of about $4.4 billion and its brands are the Levi Strauss brands, the Dockers brand, and Signature by Levi Strauss and Company, which are sold in Target and Walmart.

The company is privately held by the descendents of Levi Strauss and remains a values-based company drawing on the values of its founder, which are empathy, originality, integrity, and courage. Levi Strauss started the tradition of corporate responsibility within the company when he founded it 156 years ago. As an immigrant in San Francisco as it was growing and developing, he was already giving grants, providing philanthropy in the community as a small, new businessman. Building on that example, the company continued a tradition of philanthropy and engaging with the local community. When the 1906 San Francisco earthquake hit, the company's headquarters were destroyed, but there was a commitment to keep the employees on, and the company continued to pay all the employees and put them to work refurbishing the floor in the local facility just to keep them busy and occupied and on the payroll as the company rebuilt its headquarters. In the 1940s, Levi Strauss's racially integrated its sewing facilities in the United States. In the 1980s the company was the first multinational company to address HIV/AIDS in the workplace. The company has made a commitment to provide prevention, care, and treatment to all employees globally.

Levi Strauss is the world's largest brand-name apparel maker with sales in more than 110 countries around the world.[10] Combining direct factories with its licensed factories, there are about 800 factories in 50 countries. Levi Strauss was the first multinational company to start a code of conduct, in 1991.[11] The code was actually motivated internally by the company's employees. Levi Strauss's publishes its factory list in an effort to be transparent,[12] and has been an advocate since 2000 for the linkage of trade and labor in U.S. trade agreements.

---

[10] More information on the company, *http://www.levistrauss.com/Company/*.
[11] "Levi Strauss & Co. Global Sourcing and Operating Guidelines," *http://www.levistrauss.com/Downloads/CitizenshipCodeOfConduct.pdf*.
[12] *http://www.levistrauss.com/Downloads/FactoryList.pdf*.

> **Box 4-1**
> **Levi Strauss & Co. Global Sourcing and Operating Guidelines**
>
> <u>Child Labor</u>
> Use of child labor is not permissible. Workers can be no less than 15 years of age and not younger than the compulsory age to be in school. We will not utilize partners who use child labor in any of their facilities. We support the development of legitimate workplace apprenticeship programs for the educational benefit of younger people.
>
> <u>Prison Labor/Forced Labor</u>
> We will not utilize prison or forced labor in contracting relationships in the manufacture and finishing of our products. We will not utilize or purchase materials from a business partner utilizing prison or forced labor.

Ms. Walker noted that the code of conduct outlines the labor and environmental conditions Levi's expects to see in its contract facilities around the world. It is based on the universal declaration of human rights and ILO labor conventions. The code is a living document, which has evolved over time. Levi's has a staff of 45 employees that are dedicated to the monitoring, remediation, and capacity building associated with the code and an additional group of third-party monitors supporting that team.

The child labor component within the code is consistent with the ILO's minimum age convention and convention on the worst forms of child labor. It is considered to be a zero-tolerance issue. If Levi's finds in its initial assessment of a facility that there is child labor, Levi Strauss cannot enter that facility until the situation is remediated. If Levi's has already been active in that facility and child labor is found, there is a policy of immediate remediation and attention. Levi Strauss tries to avoid withdrawing from a facility, and works with the facility to address the situation.

Ms. Walker spoke of the Levi Strauss experiences in Bangladesh, one of the poorest countries in the world, with 45 percent of the population living below the poverty line.[13] Apparel constitutes 75 percent of the country's exports and provides both direct employment and employment through businesses associated with the apparel industry.

Levi Strauss was already sourcing from Bangladesh when the code was put in place and was just rolling it out to its supply chain. When Levi's first started to implement its policies, the company found two facilities in Bangladesh that had workers under the minimum age, which is a clear violation of Levi Strauss's terms of engagement. The company set to work to remediate the situation, taking the context into account. Ms. Walker noted that Bangladesh has had child labor problems stemming from its poverty for years and it is common for children there to be working and supporting the entire family on their wages. It is common to not know someone's legitimate age. Birth certificates are not always issued and malnutrition makes individuals look younger or seem smaller than their actual age.

Levi Strauss began to work with the facilities in question on how to address this issue. The company obtained agreement from the two factories to continue to pay the underage workers their salaries and benefits while they attended school and offered them full-time jobs when they reached legal working age. Levi Strauss agreed to pay for their tuition and books and where schools or school facilities were not sufficient to support the workers moving from the factory into schools. Levi Strauss supported the opening or renting of space and hiring of teachers to move them in. In addition, Levi's changed its policy with these facilities such that all personnel they hire must present a school certificate stating that the applicant's age is 15 years or older and where this could not be acquired or there was some doubt of age, a dental examination was required as an alternative.

Levi Strauss was aided, in a sense, by an employers' association, the Bangladesh Garment Manufacturers and Exporters Association (BGMEA), which took ownership of the situation so that the issue began to be applied to more facilities in Bangladesh. The association set up a grant of approximately $1 million to provide education to girls in the factories or who had previously worked in the factories where there were underage girls

---

[13] "Case Study: Child Labor in Bangladesh" available at: *http://www.levistrauss.com/Downloads/CaseStudyBangladesh.pdf.*

found (75,000 underage girls overall). Ms. Walker noted that the remediation occurred because Levi's was looking to see if there was a problem and Levi's stepped in with an immediate practice.

Ms. Walker also spoke of multi-stakeholder initiatives in which Levi Strauss has participated. One was the Better Cotton Initiative (BCI).[14] Included in the initiative's global principles, criteria, and enabling mechanisms is the statement that the BCI promotes decent work. With reference to child labor, the BCI notes:

- There is no child labor, in accordance with ILO Convention 138.
- Exceptionally, in the case of family smallholdings, children may work on their family's farm provided that this work is not liable to damage their health, safety, well-being, education, or development, and that they are supervised by adults and given appropriate training.
- For hazardous work, the minimum age is 18 years of age

And considering forced labor:

- Employment is freely chosen: no forced or compulsory labor, including bonded or trafficked labour.[15]

Ms. Walker said that "child labor in the first tier of our supply chain is not endemic. I think it is more sloppiness. The factory is not paying attention or looking the other way, but it's not an endemic problem. That's not to say we don't have child labor elsewhere in our supply chain. Levi's has been looking very carefully at cotton, the source of our product. Ninety-five percent of what our product consists of is cotton and the child labor is certainly there. But as far as our direct supply chain goes, that is not the key issue."

Understanding the local context and identifying local stakeholders were important elements in Levi's efforts. For sustainability to occur, Ms. Walker argued, some level of local ownership is required. Concerning Levi Strauss's ability to replicate its practices, she noted that the agricultural setting (i.e., cotton) was different from manufacturing and practices differed in these different sectors. She also noted that replicability depends on local context, which is unique to location. Levi Strauss is interested in the ability to scale up practices. As the company focuses more on cotton, one of its steps has been to disclose its supplier list in order to find other partners in the industry to work with. Levi Strauss realizes that there is a lot of duplication in monitoring and in education and training with these facilities. Working with partners in industry might allow Levi's to coordinate responses or better use resources. Efforts to scale up practices would likely be facilitated by taking multistakeholder approaches. Government engagement is very important; if governments were doing their job (i.e., if they were applying the law, if they had the appropriate law in place, and were engaging with their private sector in their countries to avoid these situations), the need for businesses to act would be reduced.

---

[14] *http://www.bettercotton.org/.*
[15] BCI, Global principles, criteria, & enabling mechanisms, Version 1.0, July 2008. p. 9. Available at: *http://www.bettercotton.org/pics/BCI_GlobalPrinciples&CriteriaEnablersv1.0_EN%20_7july08.pdf.*

Fair Labor Association

Jorge Perez-Lopez, executive director of the Fair Labor Association (FLA), described the FLA and its work in various sectors.[16] FLA has worked primarily in apparel and footwear through factory audits (independent external monitoring). Mr. Perez-Lopez discussed the case in which FLA worked with Syngenta, a multinational company that had been stung by a report about child labor in their cottonseed farms in India and wanted the FLA to help them in dealing with the issue.[17] The FLA worked with Syngenta in part to see whether the model that the FLA had developed mostly for apparel and footwear factories could, with some changes, be applied to a cultural setting as well. However, in 2005 in the midst of the project, Syngenta sold its cottonseed business. Rather than stop the project FLA switched the agricultural sector focus for Syngenta from cottonseed to vegetable seeds.[18] The FLA has been working on okra, pepper and tomato seeds, for example, in three Indian states: Karnataka, Maharashtra, and Gujarat.

Mr. Perez-Lopez noted that child labor is a large concern in India. Estimates vary, but from 14 million to 60 million children from ages 7 to 14, many of them girls, are working. A lot of them are employed in vegetable seed farms. Syngenta alone contracts with over 6,000 small commercial farms, between a quarter of an acre and an acre. The farmers that Syngenta uses are actually contractually bound to Syngenta. They are producing seed crops, they are provided starter seed, and they have a contract that says their product will be bought by Syngenta.

The FLA initiated a project in this seed crop sector involving multiple stakeholders in order to be able to understand better the context and the situation in the field. The FLA then adapted existing benchmarks to the agricultural field. They met with the multistakeholder group on remediation and on identifying the key variables on which the FLA should concentrate. From this, the FLA developed a tool kit with audit instruments and other mechanisms, based on the FLA's approach to factories, but now applied to the agricultural sector.

When Syngenta partnered with the FLA, they agreed they would implement

---

[16] "Incorporated in 1999, the Fair Labor Association (FLA) is a collaborative effort of socially responsible companies, colleges and universities, and civil society organizations to improve working conditions in factories around the world. The FLA has developed a Workplace Code of Conduct, based on ILO standards, and created a practical monitoring, remediation and verification process to achieve those standards. The FLA is a brand accountability system that places the onus on companies to voluntarily achieve the FLA's labor standards in the factories manufacturing their products. Universities affiliated with the FLA ensure that the licensees supplying their licensed products manufacture or source those products from factories in which workers' rights are protected." Fair Labor Association, "About Us." Available at: *http://www.fairlabor.org/aboutus.html*. The FLA Code of Conduct states in part: "Forced Labor. There shall not be any use of forced labor, whether in the form of prison labor, indentured labor, bonded labor or otherwise. Child Labor. No person shall be employed at an age younger than 15 (or 14 where the law of the country of manufacture allows*) or younger than the age for completing compulsory education in the country of manufacture where such age is higher than 15."

[17] See Davuluri Venkateswarlu, "Child Labour and Trans-National Seed Companies in Hybrid Cotton Seed Production in Andhra Pradesh." 2003. Available at: http://www.indianet.nl/cotseed.html.

[18] For details on what the FLA did with Syngenta, see Fair Labor Association, "Syngenta Seeds" available at: *http://www.fairlabor.org/what_we_do_special_projects_d2.html*. See also Syngenta, "Tackling labor conditions on seed farms," available at: http://www.syngenta.com/en/corporate_responsibility/tackling_labor.html.

internal monitoring, and visit each of their farms. Syngenta adopted the FLA code of conduct in principle which goes beyond child labor to include other labor standards. Syngenta monitors about one-third of its 6,000 farms each year and has found small numbers of children working, as noted in their reports.[19]

Child labor is fairly well accepted culturally in the areas where Syngenta's seed farms are located—the more remote the area, the more child labor. To combat this, much awareness raising is needed to inform families of the value of education for their children. Syngenta has been using an incentive system to combat child labor, incorporating both incentives and disincentives. The incentive is that farms that do not use child labor receive a bonus for being able to supply the product, which has worked quite well. There is also a disincentive system. If, after so many times, a farmer continues to have child labor, then that farmer will be dropped.

The FLA conducts independent audits, using monitors in India to check-up on Syngenta's the internal audits. The FLA performed visits on a pilot basis at the end of 2008. Using a cluster method and random sampling, the FLA examined 60 farms and found that most of the violations were related to safety and health; however there were some findings in the area of child labor. Most of these involved the lack of age certification systems rather than children working.

Mr. Perez-Lopez pointed out that the FLA is concerned about remediation. "We think that is part of the FLA process that if a problem is found, there is an obligation to remediate the problem." Syngenta, he noted, has taken responsibility for any child labor problems that are found. One practice is child labor awareness campaigns, which are occurring in the villages. The FLA is engaged with the communities and with the farms in terms of improving the age certification system, which is challenging in India. The FLA has also encouraged Syngenta to continue to use and refine the incentive systems, which seem to work fairly well so far.

The FLA is also working on problems to rehabilitate child workers, but Mr. Perez-Lopez noted that is difficult for a single company or one individual to try to fix a problem like not having schools. The FLA expects to continue working with local stakeholders and will seek to strengthen programs and address child labor. The FLA may also work with other companies in the agricultural area.

Winrock International

Vicki Walker, program officer for Empowerment and Civic Engagement, presented the work of Winrock, a global agricultural, natural resource organization. Winrock works in such areas as empowerment and civic engagement, trafficking

---

[19] Recently, Syngenta along with Monsanto, DuPont, Bayer CropScience, BASF, and Dow AgroSciences have put forth a plan to eliminate child labor in contracted seed growing operations and other supply chain activities around the world. See CropLife International, "Leading Seed Producers Announce Plan against the Use of Child Labor in Agriculture." June 12, 2009. Available *at:* http://www.croplife.org/library/attachments/0e67e2de-9b8f-4420-a2f3-1f38611e14a8/3/20090612%20World%20Day%20against%20Child%20Labor.pdf. CropLife International, "CropLife Position on Child Labor in the Seed Supply Chain." June 12, 2009. Available at: http://www.croplife.org/library/attachments/db61f617-8500-46d5-828b-73a42275d3ab/3/CropLife%20Position%20Paper%20on%20Child%20Labor%20in%20the%20Seed%20Supply%20Chain%20.pdf.

prevention, women's rights, education, and child labor prevention.[20]

Ms. Walker, drawing on language from the ILO, noted a principle that provides a platform for Winrock's work: the number of child laborers in a country's child population is a key indicator of economic and social development.[21] Winrock views the education of children as key to development. She indicated how important it is to work within, and strengthen existing government structures.

Agriculture is the mainstay of developing countries, with over 70 percent of the economies dependent on agriculture, which occurs mostly in subsistence smallholder farms. Over 50 percent of child labor is hidden in the non-formal sector. For example, one larger category in which much of child labor is hidden is domestic work. When children work in a household or on family farms, this adds significant complexity to the situation. Children cannot be removed from family farms in the same way that they can be removed from plantations, for example.

Ms. Walker spoke of the Community-based Innovations to Reduce Child Labor through Education (CIRCLE) project, established in 2002. The project's objective is to promote and document the best practices of innovative, community-based projects that address the reduction of child labor through formal and non-formal education. Funded by the U.S. Department of Labor, CIRCLE is an outreach program consisting of 101 subprojects affecting 24,000 children and reaching 84 local NGOs in many different sectors.[22]

Ms. Walker offered two criteria for reducing child labor in small scale agriculture:

- The criteria need to correspond to and improve local realities as well as to reflect local realities. There may not be absolutes.
- There should be a framework for continuous improvement with benchmarks over time to meet standards and criteria.

Ms. Walker then gave three case studies that meet these criteria: (1) the Child Labor Alternatives through Sustainable Systems in Education (CLASSE) program, (2) the Empowering Cocoa Households through Opportunities in Education Systems (ECHOES) program, and (3) the Tanzanian Education Alternatives for Children (TEACH) program. The CLASSE program is a program that reflects the first criterion and focuses in part on the notion of where children came from to end up in child labor. The ECHOES program focuses on relevant education and remediation. Both programs focus on the cocoa sector. The TEACH program focused on monitoring and targeting of leaders and community activists. The key criterion common to all three is the engagement of the government, the community, and alternative relevant education. Features common to all three programs include a focus on child labor in small-scale agriculture, basic education and teaching relevant agriculture, and moving from child labor to a new generation of rural entrepreneurs.

The CLASSE program focuses on Mali and Cote d'Ivoire. Children regularly

---

[20] See: *http://www.winrock.org/index.asp*.
[21] ILO, A future without child labour. Global Report under the Follow-up to the ILO Declaration on Fundamental Principles and Rights at Work 2002, International Labour Conference 90th Session 2002 Report I (B)
[22] See: Winrock International, "CIRCLE," available at: *http://circle.winrock.org/*.

travel from Mali to Cote d'Ivoire to work on cocoa farms or for domestic labor. The objective of the program has been to reduce the supply and demand of the seasonal migration of children to cocoa farms by providing training and educational alternatives for children, youth, and families. The Mali program was one of prevention and focused on increased capacity of school management committees, mentoring programs, skills development of out-of-school youth, and retention in formal school. Two key areas were skills development and vocational education. During the three years that Winrock was in Mali and Cote d'Ivoire, none of the children it had targeted left the towns or villages. They did not leave for seasonal agricultural work. They stayed at home, because they had another alternative and they saw that they had a means of livelihood in their own rural environments. The CLASSE program in Cote d'Ivoire had similar results. Efforts included children and youth completing agricultural training, student scholarships, microloans to mothers, youth sensitization to HIV/AIDS and child labor, agriculture clubs, cocoa plots, school gardens, a tree nursery, and renovation of schools.

The ECHOES program focuses on the next generation of cocoa farmers. Many of the objectives indirectly affect children; other activities directly target children. Ms. Walker noted that it was important that industry has galvanized and come together with other donors.[23] The focus is on basic education, youth livelihoods, and innovative conditional scholarships. Conditional scholarships are a type of credit to mothers and sometimes to fathers, who start a private enterprise and are able to pay back the loan to the school and continue to educate their other children and develop their enterprise. It is a type of loan where there is an enterprise that develops from it. This has been a very popular, very successful program, again, a type of best practice that could be applied in more places. The livelihoods model looks at training and educational alternatives for children and youth, and is designed to improve livelihoods in the long term.

The TEACH program, also funded by the U.S. Department of Labor, is applied in Tanzania It focuses on children who work in such areas as cotton, tobacco, and animal herding. Again, the program provides educational alternatives to children. The program focuses on children in rural areas where there were small farmholders. Activities include agriculture vocational programs and support for non-formal primary schools.

Ms. Walker concluded her talk by noting different attributes that businesses bring to these endeavors, including:

- Investing in the local economy
- Offering microcredit, conditional loans, and technology
- Matching or leveraging other support
- Leading networks of partners
- Offering innovations and new ideas and
- Increasing the likelihood of sustainability

Businesses should identify reliable NGOs on the ground to implement and monitor; leverage resources and attract partners; and engage governments and other stakeholders.

---

[23] Across Cote d'Ivoire and Ghana, partners in the project included Fazer Finland, ED&F Man, Starbucks Coffee Company, Mars Inc., Olam International, Hershey, Kraft Foods, the Norwegian Association of Chocolate Manufacturers, and the Ministries of Education, Labor, and Agriculture.

International Labor Rights Forum

Bama Athreya, executive director of the International Labor Rights Forum (ILRF), shared a case study on a Japanese company, Firestone, a wholly-owned subsidiary of Bridgestone since 1988. The story begins in 1926 when Firestone bought approximately one million hectares of land in Liberia, which would become the world's largest rubber plantation. According to Ms. Athreya, back in the 1920s when the land was first acquired, Firestone compelled villagers at gunpoint to clear the land to plant the rubber trees. They were forced to clear the land and for the intervening decades have been forced to labor on the Firestone plantation. She noted that more recently, as evidenced by visits to Liberia, families were compelled to bring all of their children to work with them every day to tap the rubber trees to meet impossible quotas. For decades, then, the plantation had widespread child labor.

There was a significant and dramatic change on this plantation in 2008, however, as a result of continued international pressure and the valiant efforts of workers on the plantation itself. The adult workers organized an independent union and, for the first time in the history of this plantation, negotiated successfully a collective bargaining agreement with Firestone. That collective bargaining agreement reduced the quotas for adult tappers, contained a clause prohibiting child labor, and raised the wages for adult laborers so they could afford to feed their kids and send them to school as opposed to bringing them to work every day. There is much to be done to ensure that the collective bargaining agreement is fairly implemented. However, this is a dramatic and sustainable step forward for conditions on the plantation for workers. It will not require a lengthy international campaign or a vast deployment of development resources to sustain this gain.

Over the years the ILRF has been organizing letter writing campaigns and other actions against Firestone. The company's response to the ILRF's most recent round of letter writing starts off with Firestone proudly declaring that they have recently negotiated a historic collective bargaining agreement with the Firestone Autonomous Independent Workers' Union and that agreement, again, diminishes the quota and calls for an end to child labor.

International Labor Organization

Benjamin Smith from the International Programme on the Elimination of Child Labour (IPEC) at the ILO spoke about the importance of engaging multilateral stakeholders and ILO initiatives.[24] He focused on two cases: the Banana and Flower Social Forums in Ecuador and Soccer Ball Production in Pakistan.[25]

---

[24] "The ILO's International Programme on the Elimination of Child Labour (IPEC) was created in 1992 with the overall goal of the progressive elimination of child labour, which was to be achieved through strengthening the capacity of countries to deal with the problem and promoting a worldwide movement to combat child labour. IPEC currently has operations in 88 countries, with an annual expenditure on technical cooperation projects that reached over US$61 million in 2008. It is the largest programme of its kind globally and the biggest single operational programme of the ILO." See ILO, "The Programme" available at: *http://www.ilo.org/ipec/programme/lang--en/index.htm.*

[25] He also mentioned a third tripartite initiative: the Eliminating Child Labor in Tobacco-growing Foundation (ECLT), which supports and funds local and community-based projects that combat child labor

His first example focused on the Banana and Flower Social Forums in Ecuador. These were tripartite initiatives to eliminate child labor based on social dialogue with effective trade union participation. He noted that trade unions are uniquely qualified to continually monitor child labor, which is more advantageous than periodic audits or other tools that may be able to diagnose the problem but do little to solve the underlying factors. Trade unions also have a clear incentive both to preserve jobs for adults and to make sure that their own children and the children in their communities are not exposed to child labor.

The Social Forum for the Banana Production Sector was established by the banana industry in 2003, in response to a Human Rights Watch report on child labor in Ecuador's banana plantations.[26] The report resulted in international pressure for banana certification.[27] In May 2004 the Banana Sector Plan for the elimination of child labor was launched, a key accomplishment of the banana forum.

An ILO report in 2000 noted that child labor was also a problem in the flower industry.[28] The Flower Social Forum is "an interagency working group made up of government agencies, industry associations, trade unions, nonprofit organizations, and international groups such as the ILO and UNICEF."[29] In January 2009, the Flower Social Forum was expanded to include other fundamental principles and rights at work. Mr. Smith observed, however, that the banana forum had lost some momentum compared with the flower forum.

The social forums in Ecuador also contributed to greater demand for effective government response. Twenty-six child labor inspectors were hired as part of a national plan to eradicate child labor. In this example, the government has been stimulated or encouraged to do a better job about fulfilling its duty to protect children from child labor and companies and trade unions are playing their roles according to their competencies in what they should be doing in the system. The ILO's participation in the Banana and Flower Social Forums has helped reduce tension among banana producers, exporters, and workers and to build consensus among the groups for the need to join forces to eliminate

---

in tobacco-growing and encourages and funds independent research to produce an objective picture of the conditions and level of child labour in tobacco growing; as well as establishes and shares best practices. The nonprofit organization, established in 2002, is supported by the ILO. Anonymous, "International Business Forum on Engaging Business--Addressing Child Labor: Case Studies." February 25, 2009. Available at: *http://www.ioe-emp.org/fileadmin/user_upload/documents_pdf/international_level/child_labour/atlanta_casestudies.pdf*. Mr. Smith noted that the role of trade unions were an invaluable asset to the ECLT initiative. The perspective and the credibility of having a trade union representative on the board of the ECLT really makes it stand out.

[26] For the report, see Carol Pier, "Tainted Harvest: Child Labor and Obstacles to Organizing on Ecuador's Banana Plantations." New York: Human Rights Watch , April 2002.

[27] Anonymous, "International Business Forum on Engaging Business--Addressing Child Labor: Case Studies." February 25, 2009. Available at: *http://www.ioe-emp.org/fileadmin/user_upload/documents_pdf/international_level/child_labour/atlanta_casestudies.pdf*.

[28] Cecilia Castelnuovo, Andrea Castelnuovo, Jorge Oviedo, and Ximena Santacruz. "Ecuador: Child Labour in Flower Plantations: A Rapid Assessment." Geneva, ILO: April 2000. See also: Fundation Salud Ambiente y Desarrollo, "Executive Summary: Baseline for the prevention and gradual elimination of child labour in the flower industry in the districts of Cayambe and Pedro Moncayo, Ecuador." Geneva, ILO: October 2002.

[29] Barbara J. Fraser, "Off the flower plantations and into school: agencies team with industry in Ecuador to combat child labor." National Catholic Reporter, May 25, 2007.

child labor. Some of the activities carried out by the forum to date include:

- Carrying out a series of awareness raising activities on child labor for trade unions, entrepreneurs, families, and children;
- Setting up child labor inspection and monitoring systems; and
- Playing an important role in getting banana companies to agree to labor inspections on their farms and plantations.

Mr. Smith then offered four characteristics of the two forums:

1. Effective dialogue concerning child labor, producing consensus on thorny issues such as hours and conditions of work for adolescents above the minimum age;
2. Private sector initiatives to improve family and community living standards in banana and flower producing areas and an increased emphasis on social responsibility generally;
3. Holistic, integrated approach that addresses root causes and provides viable alternatives to child labor: education and other basic social services; and
4. Labeling programs referencing the elimination of child labor.

Mr. Smith then discussed the case of soccer ball production in Pakistan.[30] The trigger for action, he noted, was media exposure of child labor in major sporting goods companies' supply chains, which occurred in the mid-1990s.[31] A result of this attention was that in 1997, the Atlanta Agreement was signed with the World Federation of Sporting Goods, ILO, UNICEF, and the Sialkot Chamber of Commerce and Industry.[32] With U.S. Department of Labor and other donor support, IPEC began a project there that centralized production and set up a child labor monitoring system.[33] There was a transition to an Independent Monitoring Association for Child Labor (IMAC) in 2003, as the ILO-IPEC project was in its final stages.[34]

According to Mr. Smith, one interesting element about this case was that changes to the business model (e.g., moving to centralized stitching centers) were necessary and effective in this case. Getting rid of household production, which is just inherently conducive to child labor, was important but so too was protecting women's access to

---

[30] For background on this case, see: Anonymous, "International Business Forum on Engaging Business - Addressing Child Labor: Case Studies." February 25, 2009. Available at: *http://www.ioe-emp.org/fileadmin/user_upload/documents_pdf/international_level/child_labour/atlanta_casestudies.pdf*.

[31] The majority of all soccer balls are made in Pakistan in Sialkot. See: Barbara Crossette, "Soccer Balls Sustain Pakistan Town." New York Times, October 8, 1990. U.S. Department of Labor, By the Sweat & Toil of Children (Volume IV) *Consumer Labels and Child Labor. 1997.*

[32] For the text of the agreement, see ILO, "ILO Partnership to eliminate child labor in the soccer ball industry in Pakistan," available at: *http://actrav.itcilo.org/actrav-english/telearn/global/ilo/guide/ilosoc.htm#Text%20of%20the%20agreement*

[33] For a description of the original project elements (Prevention and Monitoring Program and Social Protection Program), see: ILO, "ILO Partnership to eliminate child labor in the soccer ball industry in Pakistan," available at: *http://actrav.itcilo.org/actrav-english/telearn/global/ilo/guide/ilosoc.htm#Text%20of%20the%20agreement*. See also, ILO, "Elimination of Child Labor in the Soccer Ball Industry in Sialkot" available at: *http://www.ilo.org/public/english/region/asro/newdelhi/ipec/responses/pakistan/p1.htm*.

[34] For the IMAC Web site, see: *http://www.imacpak.org/*.

jobs. A solution was crafted by providing day care at the stitching centers so that women could continue to work and have safe alternatives for childcare. An emphasis on capacity building, as opposed to sanctions and the termination of contracts, was also employed. Overall, efforts were anchored in a broader social service platform with stakeholders contributing each according to their responsibility and competence.

Once the transition was made from an ILO-supported project to a completely industry-supported initiative, Mr. Smith noted, it was a real test for the Sialkot model whether this was going to be sustainable, but it continues to this day. In order for the local manufacturers to be able to say they are part of IMAC, they have to devote financial resources to a monitoring system that is independently checked by IMAC. The huge majority of soccer ball producers in Sialkot participate in IMAC. The risk that they will be implicated in child labor is really reduced for all international buyers because all of the producers are involved even though international buyers only work with a select few.

The soccer ball production project also brought about increased social services that were key to addressing the root causes of child labor. These have been maintained to some extent, according to Mr. Smith. "I think there has probably been some decline since the project left and the project stopped mobilizing local governments and the direct provision of education and social services, but it certainly has been improved and is a stronger social safety net in place."

Mr. Smith concluded by noting the issue of different actors: international brands and local suppliers. The international brands have monitoring systems and conduct social audits, which can diagnose the problem, but they do little to address root causes of child labor in Sialkot. The IMAC and the community-based monitoring system create a broader alliance geared to providing and ensuring real alternatives to child labor and providing education and health and the basic social services that can address the root causes. There's a disconnect between the two and Mr. Smith thought certainly that it would be in a brand's interest to be engaged with the IMAC process because of the good platform that it provides and its ability to mobilize the different actors, labor inspectors, local government, schools, and trade unions. IMAC probably would benefit from better engagement with the brands.

Ms. Aurelie Hauchere from the ILO's Special Action Program to Combat Forced Labor—a technical cooperation program that provides assistance to the ILO's constituents—then spoke about the ILO's work on forced labor. Ms. Hauchere began by defining forced labor based on ILO Convention No. 29. She emphasized that the definition focused on "all work or service that is exacted from any person under the menace of any penalty for which the said person has not offered himself voluntarily." She noted that while the definition was the base for ILO activities and programs, it needed to be translated into operational terms in practice.

Ms. Hauchere then spoke to ILO efforts with the private sector and business. She noted that the ILO has carried out a series of activities, including:

- awareness raising
- policy development (supporting the design of codes of conduct)
- identifaction of 10 principles for business leaders to combat slavery and trafficking
- training and capacity building; and

- tools and guidelines, in particular, a handbook for employers that the ILO has produced and published, which is being translated into several languages and includes some useful case studies.[35]

Ms. Hauchere noted several general challenges facing the private sector. How far should a company's liability extend? She noted that supply chains are very complex. On the one hand, there is agreement that addressing just the first tier is not enough because as far as forced labor is concerned it mainly occurs in the informal economy as is the case with child labor. But on the other hand, it is not realistic and it is not feasible to ask a company to address the whole supply chain. The question, she suggested, is how do you detect forced labor in those tiers?

Ms. Hauchere then discussed the example of Brazil as a very good practice of a multi-stakeholder initiative, which focused on rural forced labor—bonded labor—mostly in the larger Amazonian states. Forced labor is a problem in Brazil. The ILO describes the situation: "Forced labor practices in Brazil occur in rural and urban areas mainly through debt bondage schemes. In rural areas workers are immobilized in estates until they can pay off debts often fraudulently incurred; their identity documents and work permits are frequently retained; they are often physically threatened and punished by armed guards and some have been killed while attempting to flee. Debt bondage involves abusive labor contracting schemes operated by contractors known locally as *empreiteiros* or *gatos*, often engaged in other types of seasonal labor contracts. The typical debt bondage cycle occurs as follows: given the seasonality of rural demand for labor, the *gatos* recruit workers from poverty stricken areas marked by seasonal unemployment or drought. They are ferried in trucks or buses to destination sites hundreds or thousands of kilometres away from their origin. Even before they start working they will have incurred debts in initial transport and food payment at prices beyond their control. Once working they will then incur additional debts in tools, housing and other services often through abusive charges."[36]

In Brazil prior to 1995 the government denied the existence of forced labor. This situation changed that year, as the government officially recognized the problem at the United Nations. The first step was the creation of a Special Mobile Inspection Group, which is a labor inspection team consisting not only of labor inspectors but also labor prosecutors and federal police officers that go to investigate complaints on the estates to see whether there is forced labor. The teams' objective is to investigate complaints of slave labor, free workers, and to prosecute the owners of estates. If there is forced labor they have the capacity to release workers.

More than 32,000 workers have been freed in Brazil between 1995 and 2008. Ms. Hauchere attributed the success of the activity to at least two factors. First, because teams have no local offices, there cannot be any corruption from local and state owners or local police officers, which is a very important point. Second, complaints of forced

---

[35] ILO Special Action Programme to Combat Forced Labour, *Combating Forced Labour: A Handbook for Employers & Business*. Geneva: ILO, 2008.
[36] ILO, "Combating Forced Labour in Brazil" available at: *http://www.ilo.org/sapfl/Projects/lang--en/WCMS_090984/index.htm*. See also Carlos Juliano Barros, "Modern-Day Slavehouse" Rolling Stone Brazil, February 2008 and the case study in ILO, Combating Forced Labour: A Handbook for Employers & Business, Vol. 7: Good Practice Case Studies. Geneva, ILO: 2008.

and child labor violations are confidential. Visits are not announced in advance. Even for the members of the team they know the destination only a couple of days in advance.

The second step taken by the government was the creation in 2004 of the *lista suja* or the "dirty list," a register of names of individuals or company employers who have been caught using forced labor during a labor inspection.[37] The list was established by the Ministry of Labor and Employment (MTE) by a decree, and 199 employers have been on it.[38] What makes this register so effective is that it is public and is accessible via the MTE Web site.

Ms. Hauchere briefly explained how the list works. When an employer is registered on this list, it is monitored for two years and in any case will remain on the list for two years. After two years, the employer's name can be removed from the list provided he has not repeated the offense (which happens), he has paid all fines, and he has paid all labor and social security compensation. If all these conditions are met the employer's name is removed. What is interesting in this example is that in the decree there are no penalties for the employer except the fact of appearing on that list, but some penalties have been imposed by public and private financial institutions. For example, the Bank of Brazil, the Bank of Amazonia, the Brazilian Development Bank, and Northeast Bank[39] decided to refuse credit and banking benefits to the employers appearing on the dirty list. These actions can provide real disincentives for using forced labor. In addition, the National Congress is currently considering amending legislation to create new penalties, such as the expropriation of land of farm owners using forced labor.

What was the response from the business sector? According to Ms. Hauchere, one positive response involved signing a voluntary commitment: the National Pact for the Eradication of Slave Labor, which was launched in May 2005.[40] Currently about 200 companies have signed on. Together these companies represent about 20 percent of the gross domestic product of Brazil. By signing this pact, companies commit not to use forced labor and not to work with companies using forced labor.

Ms. Hauchere detailed the main commitments of the Pact:
- Commercial restrictions on enterprises and individuals identified as using slave labor
- Formalizing employment relations
    - Fulfillment of all labor and social security obligations
    - Preventive actions on safety and health
- Prevention
    - Provide information to workers vulnerable to enticement into slave labor
    - Publicity campaigns to prevent slavery

---

[37] MTE Decree No. 540/2004. For text see: *http://www.mte.gov.br/legislacao/portarias/2004/p_20041015_540.pdf*.

[38] MTE, "Cadastro de Empregadores-Portaria 540 de 15 de Outubro de 2004 ATUALIZAÇÃO Semestral em 29 de Dezembrode de 2008." Available at: *http://www.mte.gov.br/trab_escravo/lista_2009_06_16.pdf*.

[39] According to the ILO: "In December 2005 the Federation of Brazilian Banks (FEBRABAN) decided to suspend credits to companies included on the Government's "laundry list". See: ILO, "Regional Meeting for the Americas 2-5 May - Forced labour in Latin America: Fighting impunity." May 2, 2006. Available at: *http://www.ilo.org/global/About_the_ILO/Media_and_public_information/Press_releases/lang--en/WCMS_069168/index.htm*.

[40] See Reporter Brasil, "Brazilian Pact to Eradicate Slave Labour" available at: *http://www.reporterbrasil.com.br/pacto/conteudo/view/9*.

- Rehabilitation of freed workers
    - Support social reintegration for workers
    - Training and professional qualification
- Support implementation of the National Plans to Eradicate Slave Labour
- Monitoring
    - Monitor the actions and publicize the results
    - Collate and share experience
    - Assess, after one year, the implementation results

She noted that a commitment like this has to be monitored because the ILO saw that some companies did sign the pact and were found later to be using forced labor. There is a monitoring process and those companies have been excluded from the pact.

According to Ms. Hauchere, business in Brazil has taken further steps in signing the pact. Some businesses have ceased trading with estates on the "dirty list." They provide training to staff responsible for buying soya to inspect the labor practices of supplying estates. They consult the list before dealing with new suppliers. They maintain restrictions on enterprises once they have been removed from the list. On June 24, 2008, the Declaration on Social Responsibility of Companies and Human Rights was signed by 13 presidents of important national companies, Brazilian branches of multinational companies, and banks. The declaration stated that they will eradicate forced labor in their supply chains.

Ms. Hauchere concluded her presentation by discussing the role of private-sector businesses involved in labor rights activities. In Brazil, Citizen's Charcoal Institute (ICC), has worked to provide formal contracts with labor rights to freed workers--workers that have been freed from forced labor. ICC is starting with small numbers: 46 freed workers received formal contracts with labor rights in 2006, and over 100 in 2007, but that is a start and that is very important. As Ms. Hauchere noted it is important to think about what happens to freed workers.

Overall, Ms. Hauchere suggested that the Brazilian experience is a good practice because it is based on a clear definition of slave labor, has strong political commitment, is a multi-stakeholder initiative, uses both "carrots" and "sticks," includes activities by the government and businesses, has the active involvement of the labor inspectorate, and can provide contracts with labor rights.

For the ILO, an effective practice would be a tripartite process. This is the ILO way of working, involving government, employers, and workers. This should be based on social dialogue. This should definitely involve labor administration and labor inspectors. There should be a strong legal framework around this with clear indicators on what is forced labor and what isnot forced labor and, of course, based on ILO core conventions.

5
CRITERIA FOR ASSESSMENT

Dr. Susan Berkowitz presented a conceptual approach to the framework for assessing practice that included both the context and criteria for identifying good practices. The context was discussed in Chapter 3. This chapter addresses the criteria for assessment.

Dr. Berkowitz presented an overview of the draft criteria that the committee developed to begin a discussion at the workshop on what criteria would be useful for DOL/ILAB to employ in developing a list of standard practices. Dr. Berkowitz noted that the draft criteria were in no way meant to be a finished product, but were a "starting point for a lot of work and discussion."

The criteria posed by the committee drew on existing literature. In particular, the committee regarded the ILO's document *Time-Bound Programme: Manual for Action Planning*, released in 2001, as a useful place to start thinking about criteria for assessing practices. Dr. Berkowitz noted that the planning committee "looked at the way this document defined good practices (see Box 5-1) and drew from it, although we did not necessarily accept it wholesale. There was a lot of discussion about which of the criteria that were laid out in this document we thought would be appropriate for our particular uses." The committee agreed on these criteria specifically because they were general enough to allow for consistent evaluation across a range of practices implemented in varying contexts.

The committee's draft criteria (Box 5-2) consisted of six concepts: impact, program/practice effectiveness, replicability, sustainability, cost effectiveness, and relevance.[1] Dr. Berkowitz discussed each of the six elements in the draft criteria. (In the handout produced for the workshop—and reproduced below—a series of questions was provided to help identify important themes within each criterion. The questions themselves are not the criteria, however.)

Impact was meant to be taken broadly, going beyond the confines of the program or the particular practice. Did the program reduce child or forced labor, assuming that was—or should have been or would be—the larger goal around any such program. Did it benefit child or forced laborers? How was impact measured? Dr. Berkowitz noted that it was important that some sort of evaluation, even if not a formal one, be applied to the practice. She noted that it was important to include both direct and indirect or unintended consequences. She noted that these latter consequences can be quite large.

---

[1] Dr. Berkowitz noted that these terms are not necessarily universal so it is very important that we define what we mean by the terms for the purposes of this workshop because there may be somewhere else were people would not use these terms this way. There is no absolute convention about the meaning of these terms.

BOX 5-1
ILO Good Practices

- Innovative or creative
    What is special about the practice that makes it of potential interest to others? Note that a practice need not be new to fit this criterion. For example, often an approach may have been in use for some time at one setting, but may not be widely known or have been applied elsewhere.
- Effectiveness/impact
    What evidence is there that the practice actually has made a difference? Can the impact of the practice be documented in some way, through a formal programme evaluation or through other means?
- Replicability
    Is this a practice that might have applicability in some way to other situations or settings? Note that a practice does not have to be copied or "cloned" to be useful to others.
- Sustainability
    Is the practice and/or its benefits likely to continue in some way, and to continue being effective, over the medium to long term? This, for example, could involve continuation of a project of activity after its initial funding is expected to expire. But it could also involve the creation of new attitudes, ways of working, mainstreaming of child labor considerations, creation of capacity, etc., that could represent legacies of a particular practice. This criterion may not apply to all types of practices.
- Relevance
    How does the practice contribute, directly or indirectly, to action of some form against child labor?
- Responsive and ethical
    Is the practice consistent with the needs, has it involved a consensus-building approach, is it respectful of the interests and desires of the participants and others, is it consistent with principles of social and professional conduct, and is it in accordance with ILO labor standards and conventions?
- Efficiency and implementation
    Were resources (human, financial, material) used in a way to maximize impact?

SOURCE: IPEC, *Good Practices: Identification, Review Structuring, Dissemination, and Application.* Geneva: ILO, October 2001, pp. 2-3.

> **BOX 5-2**
> **Criteria for Assessing Practices**
> **Draft Proposed at Workshop to Facilitate Discussion**
>
> 1. Impact
>   a. Did the program reduce child or forced labor?
>   b. Did it benefit child or forced laborers?
>   c. How was impact measured?
>   d. Direct, indirect, and unintended consequences?
> 2. Program and/or practice effectiveness
>   a. Did the program achieve its goals?
> 3. Replicability
>   a. Could the practice be implemented with modest adaptation in other settings?
>   b. What factors limit/encourage replicability?
> 4. Sustainability
>   a. Is the practice likely to continue (as needed)?
>   b. Is the benefit likely to continue effectively?
>   i. Is the institutional capacity necessary to sustain benefits or practices?
>   ii. Local ownership
> 5. Cost effectiveness
>   a. Were the benefits adequate in relation to likely benefits from comparable investments?
>   b. How is cost effectiveness measured?
> 6. Relevance
>   a. Is there a set of assumptions about how activities will lead to outcomes? (Do you understand why the program works?)
>   b. Is there a logical connection between the inputs, activities, and outcomes?

Program or practice effectiveness—the second element—was more focused on the program or intervention and asked whether it achieved its goals. It is a somewhat narrower construction than the notion of impact.

Replicability focuses on the extent to which a practice is unique to a particular situation or can be used elsewhere. What kind of modifications would be needed to have a program that is successful in one venue be successful in other venues without significant adaptation? A related question is to ask which factors limit or encourage replicability, because that helps one to understand the context in which a practice can work or not.

Sustainability is, in a sense, the other side of replicability and focuses on whether a practice can continue beyond an initial outlay of resources (e.g., initial funding). Individual programs or initiatives are often funded for a set period of time and then financing is reduced or eliminated (sometimes with the idea that other sources of funding will be found by the time the initial period ends). This is very common in almost any kind of initiative, according to Dr. Berkowitz. Then the question has to be, can it continue on in the absence of this funding? What sort of lasting infrastructure or activities did the initial funding produce? This raises the further question of institutional capacity building: has a program laid the groundwork, for example, via local ownership, so that it can continue?

Cost effectiveness was meant to focus on the balance between the infusion of resources and the outcomes of the practice. When the committee was crafting the draft criteria, Dr. Berkowitz suggested that it was concerned about practices that might work very well, but at very high cost. Businesses ought to have a methodology for measuring cost effectiveness and incorporate this concept into the strategic thinking.

Relevance focuses on the underlying theory linking the practice to its desired outcomes. Is there a set of assumptions about how activities will lead to outcomes? This came from a discussion about logic models, which are often used in program evaluation. Logic models identify the inputs, the activities, and the outcomes (short term and longer term) that are assumed to derive from that. The connections between those three elements are important. Dr. Berkowitz suggested it is important to understand the connections between these components of practices. One reason is that to judge whether a program has been effective, one needs to understand these assumptions.

Mr. Viederman (Verite and a committee member) noted that the committee believed that in order for something to be considered a good or good enough practice not all of these six elements must be present; for example, a program that has a large positive outcome and is cost effective but not replicable might still be a good practice.

Comments on the Criteria

Participants at the workshop—both presenters and audience members—reacted to the draft criteria in various ways. One presenter said they were useless and another too generic. Others found them very useful. Most offered suggestions for clarifying, adding to, or changing the criteria presented. This section begins with general comments about the criteria, followed by comments made to specific criteria. (Additionally, the National Research Council received written comments on the criteria, which have been reproduced in Appendix G.)

General Comments

1. There is no need for the criteria, because they already exist.

Bama Athreya from the ILRF suggested that there was no need for new criteria as criteria already existed: "Apparently no mapping was done of the number of excellent sets of criteria that have been developed over a period of decades now to ensure (1) that investors do not have the risk of seeing their investments placed in products or companies that use child or forced labor, (2) that the labor rights community has undertaken to see the broad applicability of core labor standards, and (3) that consumer rights groups have undertaken to ensure that consumers are not purchasing goods made by child or forced labor. I have in my folder here just one set of such criteria developed and they are criteria, let me be very clear about that, developed by a European pension fund and spelling out exactly what sorts of due diligence that fund is undertaking with regard to companies it holds to ensure that those companies are employing best practices to reduce or eliminate child labor. These efforts exist. There was no need for this committee to reinvent the wheel and the fact that they chose to do so rather than mapping and simply borrowing from such best practices is troubling."

An audience participant also expressed that the criteria were generic. "That is, [the criteria] could have been picked out of any handbook on evaluation and said, here it is. The criteria as formulated do not reflect any of the experience or wisdom that is obviously present [at the workshop]. So I do not know how one would move…in the direction of formulating criteria that somehow presupposes some understanding or knowledge of what is being dealt with, but the generic quality of the criteria seems to me really emasculates, vitiates any strength it might have as a tool."

2. Criteria could be better defined.

Jorge Perez-Lopez noted that good definitions were critical: "It is going to be a challenge to define some of these things because there are very many situations that arise and how these concepts are defined is very important." Rachel Rigby commented that "it would be very helpful for our criteria…to also be accompanied by a set of definitions."

Jeff Morgan concurred, noting that terms must be very clearly defined and understood among key stakeholders. In particular, he asked for clarity on the following terms:

- "reduce" child or forced labor
- "benefit" child or adult laborer
- withdrawal, remove, rescue
- trafficking–with or without movement

One audience member felt that the difference between "impact" and "effectiveness" was unclear. Another noted, "There are also words that are used interchangeably that are not the same. Impact and consequence are not the same." According to the participant, "usually impact is something that you expect to happen. The consequence you may or may not expect it to happen, but they are not the same. An impact is directly what happens, a consequence may be something that happens down the road, it is longer term. It depends upon how you are thinking about it and so on. I do not care how you define these words, but I think you that to define them carefully so that everybody understands what you are talking about."

3. Criteria should be reordered.

One audience member noted that the criteria are not in the right order so they do not encourage logical thinking about it, about the project or about the program and so on (see an alternative presentation of the criteria in the comments submitted to the meeting in Appendix G). In a related vein, another participant suggested that the criteria be organized chronologically, that is, some elements, like relevance, could be construed as being applied before the program, while impact happened later, followed by sustainability and replicability. Another also argued for a hierarchy, but based on which elements were most important: "Obviously what we care most about is…was it effective in reducing…child labor or forced labor. Whether it was cost effective or not sort of presupposes that we have a whole universe…of models that work and that we are going to choose among them whether this one costs less in relation to the outcome or not. As best I know, that is not the case so I think there is a need for some sort of hierarchy here. What is important? I would pick out, personally, did it work and that it is replicated several times here. Maybe can it be replicated?"

4. Greater clarity is needed as to how many criteria a practice has to meet to qualify as a "good" practice.

As Dan Viederman noted, there was not an expectation when the draft criteria were created that a practice had to meet all of them to qualify as good. However, this leaves open the question of how many criteria a good practice should meet. Rachel Rigby similarly added: "There are six criteria here. There are some subcriteria and I am sure there is no definitive answer, but exactly how many need to be present in order for it to be a good practice. You mentioned that it did not need to be all. I can think of some programs where they maybe have one, but the others are sort of weak and then I do not know if I would call that a good practice."

Jeff Morgan brought up the question of how practices would be scored or how elements would be weighted. This comment could be seen as addressing how much each criterion is counted among the six proposed, how practices would be scored in meeting each criterion or how observers would score practices. He asked for example, "What determines being placed on the DOL list or removed from the list?"

5. Context needs to be further emphasized.

Jeff Morgan argued that the criteria appear to be focused on remediation efforts (programs) only. Mr. Morgan thought the criteria were too program or project focused. He was concerned that the criteria do not consider the other elements that enable the necessary programs. The criteria, in his view, need to look at the enabling environment that would allow projects to take place. As he outlined it in his presentation:

- The criteria lack context of the environment in which the labor issue occurs.
    - What are the key issues faced by those who seek to address the problem?
        - Overall recognition, commitment, and capacity to address the issue
    - What laws and regulations are in place and enforced in the area where issues are occurring?
        - Legal, regulatory, and judicial framework
    - What is the stability of the government and the government structures in the region where the issue takes place?
        - Is the government supportive of a plan to address the issue?

Mr. Morgan mentioned that the level of recognition and type of involvement on the part of the government with regard to a problem made it more or less difficult to work with and define programs. Mr. Morgan provided the Côte d'Ivoire, the largest producer of cocoa in the world, as an example. He noted that the stability of that government has been in question since 2001 and it is still creating a lot of issues for industry when it comes to putting programs on the ground in the country.

Second, he argued that the criteria are narrow in scope and may be based on erroneous assumptions regarding the scope and root causes of the problem. He pointed

out differences he perceived between:
- "Hired" labor vs. family labor
- Agriculture vs. manufacturing
- Subsistence vs. commercial agriculture
- Economic status of families in the sector
- Status of infrastructure: schools, roads, health centers, education, communication capabilities.

Mr. Morgan said it was important to distinguish between employment activities, for example, between factories hiring laborers and family labor. He also argued it was important to distinguish between different types of agriculture and between agriculture and manufacturing. He noted that he had met farmers whose children had to work or the family would go hungry. He also noted the issue that the preferred alternative, that is, school, might not be available.

Toni Dembski offered that there was a need for understanding the objective of a business entering into one of these programs, that is, the objectives could be broad in a number of respects. She also offered that a contextual element is the scope of the importing that might be undertaken by a company. In the case of Target, its breadth—3,000-4,000 different contractors and perhaps up to 10,000 different factories—is a complicating factor.

An audience participant suggested that the criteria be grounded in three kinds of contexts. First, there is the environment in which the agent of child's work, the parent or family, makes decisions about a child engaging in work that benefits the child and the family. It is very hard to come in and impose judgment from the outside about whether we should try to defeat that or work with that and so on.

Second, there is the legal environment, which may include laws against child or forced labor or some sort of exploitation, which the participant took to be the focus of this meeting. It is important to note that the laws and the ILO Conventions do not bind farmers, they bind governments, and the government may simply lack the regulatory authority force to put the laws into affect.

In a third environment the government itself is the culprit. The participant pointed out Turkmenistan and Uzbekistan as primary examples of this. In cases such as these there is a nexus. "Democracy has been mentioned, freedom of association, but invariably child labor in these environments is imbedded in a social and political context where you just cannot go in and solve the problem of child labor. You have journalists in jail or journalists being shot. You have political activists or trade unionists who are in jail or who were suppressed or whose organizations are managed by government. To talk about child labor in this environment without placing it in a larger…political context in which these abuses occur…is really to live in a never- never land. … It seems to me if we are going to talk about child labor in that third context, we really need to talk about and address all of the other problems here."

6. Other components could be added to the criteria.

Bill Guyton urged that the use of partnerships could be a criterion. He argued that there was a need for the commitment of all stakeholders--whether government, private sector, civil society and farmers--in the case of cocoa.

Mr. Guyton also suggested that an incentive structure was a critical element that seemed to be missing from the criteria. A few audience members took up this point, noting that there was an important distinction between positive and negative incentives. One audience member summarized a few points on incentives: "I think you emphasize here regulation, compliance, and monitoring, but it is hard to regulate if the incentives are working against you and it becomes much easier if you look at the incentive structure and try to build your intervention so that some people are doing what they want to do or what is beneficial for them. I think that means first of all finding partners. In the ILO, trade unions often can be brought on board and sometimes even employer associations when they view child labor as an issue that creates an unequal playing field. Finding and using incentives also, I think, involves what I would call piggybacking, that is building your program on top of another program that is working already so you do not have to do everything yourself."

Bama Athreya noted that transparency and traceability were missing: "The criteria that we are missing or rather the subjects that seemed to be missing altogether in the development of the criteria are supply chain transparency and traceability, transparency so that consumers can have access to information about all of the suppliers in a supply chain, traceability so that companies themselves can have a system with which they can track all products back to the point of origin."

Specific comments on individual criteria

Impact

Mr. Morgan highlighted two issues with regard to assessing impact. First, he noted the goal of the Harkin-Engel Protocol: "We were asked to implement a system that would show that cocoa was being produced without any of the worst forms of child labor." He noted that if the objective is 100 percent eradication, "it is going to always be difficult to say that yes, that objective is being met." Mr. Morgan also noted that the Protocol includes deadlines that the industry has struggled to meet and that progress can be subjective.

Thea Lee suggested more attention be paid to the notions of direct, indirect, and unintended consequences, "because certainly with respect to child labor, the unintended consequences are an enormous issue for folks evaluating child labor programs. Do you send the children begging or other even worse outcomes, prostitution and so on, if you take them out of the apparel supply chain and so it seemed like in the criteria, you would want very explicitly to ask whether the program design incorporates elements to directly mitigate the unintended consequences and evaluate those elements of the program separately and give it the attention that is deserves."

Bill Guyton spoke to the issue of practices within the context of preexisting programs. He asked, "How do you look at labor criteria fitting into existing programs or preexisting programs that you may have already been operating and how do you begin to measure practical inputs and outputs for those programs?"

One audience member spoke about the difference between qualitative assessment and quantitative assessment. "It is one thing to know how many students are enrolled in school. That is not the same thing as how many students are able to attend school. Also, what does it mean for somebody to complete their education? Are there exams? Is there a measure of quality or is it just you enroll and you show up for x percentage of the year and therefore you get passed on to the next grade and so on."

Program and/or practice effectiveness

The main comment on this issue was that it was not clear what effectiveness meant. As noted above, defining terms is critical.

Replicability

One audience member noted that there is a difference between a practice being replicable and a practice being scaled up.

Sustainability

Jeff Morgan noted with respect to sustainability that the effort may be more than an industry by itself can promise. He argued that sustainability requires capacity building within the country, sustainable sources of funding, and a long-term commitment. Mr. Morgan noted that sustainability is a big question when you talk about a large industry and a large sector. He argued that it depends not only on the industry's interest, but on the interest of the government as well. Bill Guyton asked, "How do you know over time if a program is truly going to be sustainable if it will start and end and then evaporate?"

Cost effectiveness

Toni Dembski noted that it was not clear what cost effectiveness means and how it is measured. For example, is the practice cost effective for the business that is operating it or is it cost effective in some other context?

One audience member noted: How do you measure it? I get really concerned with the word "cost" because usually the cost we talk about is how many dollars or euros or whatever currency we are talking about could cost. When we talk about cost effectiveness, though, we also have to take a look at the cost to the children if the program is not in place and the cost to the forced laborers if the program is not in place. So when we use the word cost, what is it we are talking about? Are we talking about just the financial cost or are we also talking about the human cost when something like this does not take place?"

Relevance

Jorge Perez-Lopez questioned whether relevance needed to be a criterion: "I am still struggling with this issue of relevance and whether it is a criterion or whether this is just a factor or a methodological issue. I do not see how relevance really is or should be a criterion in your work."

An audience member noted that it is really of critical importance to get the logic and the logic model explicit and suggested moving that up to the beginning and to recognize that there are many logic models for child labor, that is, there are many reasons why there is child labor. One of the pitfalls, the commenter noted, in the design of child labor programs is a mismatch between what we think the problem is and what the solutions are. The audience member gave three examples to illustrate the point. One was a child labor program in Central Asia that assumed that the problem was poverty and the solution was to buy milk cows for each family in a village, so a good milk cow was the intervention based on the notion that children were working because the parents were not able to earn adequate income. A second program, in Russia, was aimed at dysfunctional families, providing family counseling for families with alcoholic parents and so forth. The assumption here being that the problem was a family problem, more sort of sociology. Third, in Central Asia too often migration is a cause of child labor. When people migrate to pick cotton or some other product, they take their children with them and there are no schools around. "So it really depends on the logic model and getting that right. I think that is just a critical first step."

6
WRAP-UP

At the end of the workshop, the committee members were each invited to give their own thoughts as to what broad themes they heard during the workshop. Susan Berkowitz began by offering her summary of the main points raised over the course of the two-day discussion:

- *Context.* The role of context (e.g., role of government, trade and economic policies) and its effect on businesses' ability to influence the situation is important. Dr. Berkowitz noted that the context is never going to be the same, so this variance results in differing abilities of businesses to addresses concerns in specific situations. Some things are going to be more amenable to change than others and it is important to have an assessment of the situation.
- *Logic Model.* Dr. Berkowitz noted that the discussion focused in part on what it would mean to examine assumptions and have a logic model and a set of expectations about why certain outcomes would follow from certain activities.
- *Partnerships.* A strategic approach to partnering depending on the setting and the context should be considered, including focusing on multistakeholder efforts and trying to piggyback on other efforts. She noted that the discussion highlighted businesses' thinking about who could be partners, engaging with them, obtaining buy-in, and recognizing that some potential partners might not be amenable to solving the problem. Partnering seemed to Dr. Berkowitz to be an important element of the discussion.
- *Incentives.* Many participants had raised the importance of incorporating incentives, both positive and negative, into practices.
- *Scalability.* Clarifying the notion of scalability, specifically whether one should view scalability as a form of replicability or as a different set of issues. An important issue was to examine how more focused or limited practices might scale up to the national level and to what extent they would work as they got larger.
- *Data Collection.* Having some kind of data collection that has an assessment or an evaluative component to it as part of a process of both the documenting and assessing practices.
- *Knowledge Sharing.* Once practices have been assessed, it is important that knowledge of both what has worked (good practices) and what has not worked be disseminated to those working in this area so that they can benefit from the experiences of others, avoid "reinventing the wheel," and avoid practices that do not work.

Beryl Levinger also noted the importance of context. She noted that many talks at the workshop focused a lot of attention on this and she felt that one could distill a preliminary framework for contextualization. According to Dr. Levinger, "Contextualization in any criteria has to be closely linked to the environment in which the labor issues occur. This environment includes supply chain characteristics and the degree of government support for labor regulations and enforcement; government capacity to deal with labor issues; the size and scope of the sector; the composition of the sector; the role, engagement, and involvement of industry in the sector; and the role of families and communities in the sector. Without a fairly comprehensive understanding it is difficult to propose criteria that are contextually relevant."

A second comment from Dr. Levinger had to do with the underlying assumptions. Whatever criteria are ultimately developed should reveal underlying assumptions about what makes business practices effective. "The assumptions that we may or may not be making have to do with whether we are talking about hired labor versus family labor, whether we are talking primarily about agriculture or manufacturing. If we are talking about agriculture it matters whether it is subsistence or commercial. Also, what assumptions do we make about the nature of the economic well-being of families in the sector, and finally status of the infrastructure, including schools, roads, health facilities, and education institutions?"

Dr. Levinger emphasized the importance of school enrollment and attendance rates for primary school education as a predictor and an important contextual element. Kevin Bales interjected here that measures of education were significant predictors of labor problems, but not quite as good as indicators of corruption or the rule of law. Dr. Levinger noted that these together might be useful. Dr. Bales agreed and added indicators of social unrest and conflict. He noted that including a measure for poverty created a good model for predicting labor issues. Dr. Levinger noted that such a model might be useful because it forms the situation in which business practices are condcuted.

Dan Viederman added that one theme which came out very strongly in the context of business practices in particular was integration. Business practices, he suggested, are probably better to the extent that the core business practice, the sourcing, the buy-in practice are integrated with the responsibility practices. Business practices are probably more effective as they address forced and child labor to the extent that the responsibility aspect of the business or orientation of the business is integrated with whatever else the business is doing on public policy terms, be it lobbying explicitly or decision making or the approach that the business takes to issues like corruption. He posited that there is an overlap in being able to identify, find, and fix forced and child labor if one is also identifying, finding, and fixing other core labor standards, freedom of association in particular. Integration as a criterion against which to measure business practices such that we can predict or assume their effectiveness is an important one to include.

Adam Greene was struck by the discussion of what he called "success factors," those features that would determine whether a business practice was effective or not. These features may include stakeholder engagement and pulling in, coordinating with other groups, and mobilizing different actors in their different roles. A business practice may or may not be effective depending on whether it is one organization by itself or part of a coordinated effort in different actors.

Mr. Greene noted a related point was the link between child labor and forced labor and the other core labor standards. He noted that forced labor rarely happens by itself. He suggested that the criteria should take into account how child and forced labor relate to the other core labor standards and how you would use that as leverage to try to address the problem, or as he put it "a holistic, integrated approach being a success factor." He also noted the importance of understanding and addressing root causes.

Mr. Greene commented on the context, arguing that a practice in one environment may or may not have the same impact in another operating environment. He concluded: "That's a piece that's going to have to be considered going forward. Again, it may be something that has to be considered as part of business practices but it is the success factors that will have a big impact."

Donna Chung noted that she agreed with the comments already stated by the other committee members, and wanted to add one point on the importance of clarification of terms. She noted it was critical to clarify terms in the criteria as well as other terms used during the workshop (referring to business practices, supply chains, context, etc.).

Dr. Chung also focused on the notion of credibility. She noted that when one discussed possible success stories, one needed to put them in context in terms of what was successful and according to whom (that is, the evaluation and the evaluator).

Dr. Chung raised the larger question as to the responsibilities companies have in affecting the root causes in the larger context in which child labor and forced labor take place. Kevin Bales noted in response to this point that a theme in a number of the presentations of good practices touched on the philosophical or normative or moral orientation of decision makers on corporate boards or in business, which he noted was very difficult to measure. Dr. Bales noted that this orientation was important in the decisions that companies make, so that they are focused on protecting individual rights.

At this point in the workshop, participants were invited by Dr. Berkowitz to comment on the broad themes. Again, participants focused on the need to understand the context, as for example in this comment: "One is that I still have a fear that everything is going to be lumped together when you are looking at the criteria and that there is such a stark difference when you are looking at small-scale agriculture, family run, family operated versus a wage laborer in a manufacturing setting, and it's not at all the same. I think it is important to make that clear distinction and it might be that the criteria are different depending upon the situation."

Another audience member commented on the point made by Kevin Bales regarding businesses' moral compass. The audience member noted that faith-based investors and socially responsible investment managers focused on pressuring companies that did not do it on their own to adopt codes of conduct, set up compliance systems that are monitored, and broadly disseminate information on the results of efforts and monitoring. The audience member noted the important role that shareholders—from the NGO community, the faith-based community, and the socially responsible investing community—have had and continue to have.

APPENDIXES

Appendix A
Committee Member Biographies

Susan Berkowitz (Chair), a Senior Study Director at Westat, is a recognized expert in mixed method evaluation design and implementation as well as analysis of qualitative data who currently chairs Westat's Evaluation Working Group. During her nearly 20-year career at Westat, Dr. Berkowitz has led studies of higher education policies and programs, military and civilian health services and health delivery systems, clinical research networks, patient participation in medical decision making, health communication campaigns, public-private partnerships, youth career decision-making and military propensity and interventions aimed at high-risk, abused and neglected children and frail, low income elderly. She has delivered technical assistance in evaluation design and implementation to National Science Foundation and Office of Special Education grantees and given training and professional development workshops to a variety of audiences. Dr. Berkowitz wrote a widely cited chapter on qualitative data analysis for The User-Friendly Guide to Mixed Method Evaluations (NSF: 1997) and co-edited and co-authored Needs Assessment: A Creative and Practical Guide for Social Scientists (Taylor and Francis: 1996). She is a frequent presenter at national and international meetings.

Kevin Bales is president of Free the Slaves (www.freetheslaves.net), the United States sister organization of Anti-Slavery International, and Professor Emeritus of Sociology at Roehampton University London and Visiting Professor at the Wilberforce Institute for the Study of Slavery and Emancipation, University of Hull. His book *Disposable People: New Slavery in the Global Economy* was nominated for a Pulitzer Prize, and published in 10 languages. Desmond Tutu called it "a well researched, scholarly and deeply disturbing expose of modern slavery." In 2008, Utne Reader named Dr. Bales as one of "50 visionaries who are changing your world." In 2006 his work was named one of the top "100 World-Changing Discoveries" by the Association of British Universities. He won the Viareggio Prize for services to humanity in 2000. The film based on *Disposable People*, which he co-wrote, won a Peabody Award and two Emmy Awards. He was awarded the Laura Smith Davenport Human Rights Award in 2005; the Judith Sargeant Murray Award for Human Rights in 2004; and the Human Rights Award of the University of Alberta in 2003. He was a consultant to the United Nations Global Program on Human Trafficking. Dr. Bales has advised the United States, British, Irish, Norwegian, and Nepali governments, as well as the ECOWAS Community, on slavery and human trafficking policy. In 2008 he was invited to address the Summit of Nobel Peace Laureates in Paris. He is currently writing a book on the relationship of slavery and environmental destruction; and with Jody Sarich, a book exploring forced marriage worldwide. He earned his Ph.D. at the London School of Economics.

Donna E. Chung is a Trade and Labor Compliance Advisor at Sandler, Travis & Rosenberg, P.A., and heads the firm's corporate social responsibility practice. Before joining ST&R, Dr. Chung worked in the U.S. Department of Labor's Office of Child Labor, Forced Labor and Human Trafficking, managing child labor elimination projects in Latin America and the Caribbean. Dr. Chung's doctoral research focused on international labor standards and CSR in China's apparel and footwear industries. Her research also included the creation of a framework for understanding CSR in the context of free trade agreements and U.S. trade policy. Dr. Chung serves as adjunct professor at the American University's School of International Service and teaches a master's course on international policy analysis. She also serves as advisor to SISHA, a Cambodia-based nongovernmental organization that fights human trafficking and sexual exploitation in Southeast Asia, and as a member of the National Academy of Science's Committee on Reducing Forced and Child Labor. She is fluent in Korean and has a working knowledge of Spanish. Dr. Chung has a Ph.D. and M.Phil in international relations from Oxford University and a B.A. in ethics, politics, and economics from Yale University.

Eric Edmonds joined the faculty at Dartmouth in 1999. Currently, he is an associate professor of Economics. He is director of the Child Labor Network at the Institute for the Study of Labor (IZA), a faculty research fellow at the National Bureau of Economic Research, a research fellow at IZA, and an associate editor at Economic Development and Cultural Change. His research focuses on improving our empirical understanding of the reasons for the prevalence and persistence of child labor, illiteracy, and low levels of schooling attainment in low income countries. Dr. Edmonds received his Ph.D. in Economics from Princeton University and a M.A. and B.A. in Economics from the University of Chicago.

Adam Greene is vice president, Labor Affairs and Corporate Responsibility at the U.S. Council for International Business. He is responsible for USCIB's activities on labor policy as well as our wide-ranging work on corporate responsibility. He manages U.S. business participation in the development of international labor standards and advises companies on international and regional trends in labor and employment policy. He also coordinates USCIB involvement in the governing and standard setting bodies of the International Labor Organization and promotes the ILO Declaration on Fundamental Principles and Rights at Work. He is vice chairman of the Business Technical Advisory Committee on Labor Affairs to the Inter-American Conference of Ministers of Labor. In the area of corporate responsibility, Mr. Greene advises clients on international codes and initiatives, internal management systems, strategic alliances and corporate reporting, among other things. He is a member of the ISO Working Group on Social Responsibility and Co-Chair of the Industry Stakeholder Group, and is a member of the U.S. Advisory Committee for the FTSE4Good social investment index. Mr. Greene is actively involved in the ongoing implementation of the OECD Guidelines for Multinational Enterprises, the ILO Tripartite Declaration on Multinational Enterprises and Social Policy, and a wide array of other international initiatives.

Beryl Levinger, since 1992, has held the appointment of Distinguished Professor of Nonprofit Management at the Monterey Institute where she also serves as Academic Director of the Development Project Management Institute (DPMI). Prior to joining the Institute's faculty, she held significant leadership positions in international nongovernmental organizations (NGOs): president of AFS Intercultural Programs; senior vice president of CARE; and senior adviser to Save the Children's president. From 1992 until 2007, Beryl also directed the Center for Organizational Learning and Development at the Education Development Center. Center clients include NGOs, governments, and multilateral institutions seeking to craft new strategies, develop greater capacity, or achieve organizational transformation. Beryl is a former vice chair and a founder of InterAction, the coalition of 175 U.S.-based international NGOs focused on the world's poor and most vulnerable people. She has written extensively on capacity building and development. Her books include *Togetherness: Intersectoral Partnering in Latin America*; *Critical Transitions: Human Capacity Development Across the Lifespan*; and *Nutrition, Health, and Education for All*. Since 1999, Dr. Beryl has been research director for the State of the World's Mothers Report, an annual policy-oriented publication supported by Save the Children. Dr. Beryl has been appointed to serve on a number of prestigious expert groups. Current or recent panel memberships include: Ending Child Hunger (sponsored by the World Food Programme); Standards for Educational Administrators (sponsored by the National Policy Board for Educational Administration and the Council of Chief State School Officers); and the World Bank's Development Gateway Advisory Group Beryl received her M.A. and Ph.D. in educational planning and administration from the University of Alabama. Her undergraduate degree in the social sciences was awarded by Cornell University.

Dan Viederman is executive director of VERITÉ. Since first going to China as an educator in 1985, Mr. Viederman has passionately pursued partnerships between international NGOs and domestic institutions in China and Southeast Asia. He has lived in Asia for ten of the past twenty years, guiding international nongovernmental organizations into effective support for social change and sustainable institution building. Since becoming executive director of Verité in 2004, hen has led the expansion of global capacity and the establishment of formal partnerships with NGOs and Regional Offices around the world. This structure recognizes that sustainable change requires strong local institutions, and leverages Verité's global strength and corporate relationships to support our in-country partners. Under Mr. Viederman's leadership Verité has become a recognized source for thoughtful commentary on the impacts of globalization on workers around the world. . Mr. Viederman was previously CEO of the China Program for WWF-World Wildlife Fund, where he established the Beijing office for the first international environmental NGO in China. In that role he worked extensively with government and private sector institutions to support conservation outcomes, linking on-the-ground understanding with national policy goals, and providing key strategic leadership to the first Chinese environmental and corporate responsibility NGOs. A graduate of Yale University, he has a Master's Degree in International Affairs from Columbia University and a certificate in Chinese language from Nanjing Teacher's University. He was a San Francisco Coro Foundation Fellow, and serves on the Boards of Clean Air-Cool Planet, Dwight Hall at Yale and as Chair of the Educational Opportunities Fund.

Appendix B
Workshop Agenda

Identifying Good Practices for Producers/Purchasers to Reduce the Use of Child or Forced Labor

Location: Keck Center of the National Academies
Room 100
500 Fifth Street, N.W.
Washington, DC 20001

Agenda

May 11, 2009

8:30-9:00 Breakfast in the room

9:00-9:20 Welcoming Remarks
Susan Berkowitz, Senior Study Director, Westat (Committee Chair)

9:20-10:30 Perspectives from the Sponsor
Charita Castro and Rachel Rigby, Office of Child Labor, Forced Labor, and Human Trafficking, Bureau of International Labor Affairs, U.S. Department of Labor

10:30-10:45 Break

10:45-12:00 Presentation on Draft Criteria
Susan Berkowitz

12:00-1:00 Lunch

1:00-2:45 Session I

Kevin Bales, President, Free the Slaves (Moderator)
Beryl Levinger, Distinguished Professor of Nonprofit Management, Monterey Institute of International Studies (Rapporteur)

Speakers:
Jeffrey Morgan, Director - Global Programs, Mars Inc.
Bill Guyton, President, World Cocoa Foundation
Meg Roggensack, Policy Director, Free the Slaves

2:45-3:00 Break

3:00-4:45   Session II

                Dan Viederman, Executive Director, VERITE (Moderator)
                Donna Chung, Trade and Labor Compliance Advisor at Sandler, Travis & Rosenberg, P.A. (Rapporteur)

                Speakers:
                Thea Lee, Policy Director, AFL-CIO
                Toni Dembski, Senior Counsel in the Law Department, Target
                Jorge Perez-Lopez, Executive Director, Fair Labor Association

4:45   Adjourn

May 12, 2009

8:30-9:00   Breakfast in the room

9:00-9:15   Welcome to day 2
                Susan Berkowitz, Senior Study Director, Westat (Committee Chair)

9:15-10:45   Session III

                Donna Chung, Trade and Labor Compliance Advisor at Sandler, Travis & Rosenberg, P.A. (Moderator)
                Adam Greene, Vice President, Labor Affairs and Corporate Responsibility, U.S. Council for International Business (Rapporteur)

                Speakers:
                Bama Athreya, Executive Director, International Labor Rights Forum
                Anna Walker, Manager, Government Affairs and Public Policy, Levi's
                Benjamin Smith, Chief Technical Adviser, ILO-IPEC

10:45-11:00   Break

11:00-12:30 Session IV

Beryl Levinger, Distinguished Professor of Nonprofit Management, Monterey Institute of International Studies (Moderator)
Dan Viederman, Executive Director, VERITE (Rapporteur)

Speakers:
Aurélie Hauchère, Technical Project Officer, ILO- Special Action Programme to Combat Forced Labour
Mark Neuman
Vicki Walker, Program Officer, Empowerment and Civic Engagement, Winrock International

12:30-2:00 Lunch

2:00-3:00 Summary/Closing remarks

Susan Berkowitz
Kevin Bales
Donna Chung
Adam Greene
Beryl Levinger
Dan Viederman

3:00 Adjourn

Appendix C
Speaker Biographies

Bama Athreya is the executive director for the International Labor Rights Forum, a Washington, DC-based nonprofit advocacy organization. The ILRF promotes worker rights worldwide through research, publications, public education and advocacy related to trade agreements and corporate accountability. At the ILRF, Dr. Athreya has developed new programs and advocated for stronger protections for workers' rights with governments, multinational corporations and international organizations. She developed and launched new work including the Rights for Working Women Campaign to research, understand and promote viable remedies for sexual harassment and violence in the workplace; the Ethical Garments Project working with brands, labeling initiatives and public procurement efforts to create a 'sweatfree' standard for apparel production worldwide; the China Rule of Law Project which trains judges, arbitrators, lawyers, and employees of government legal aid centers and trade unions in labor law within China; and work with partner organizations in each of the Central American countries to develop comprehensive independent assessments of national labor standards. Dr. Athreya joined the ILRF in early 1998, just after returning from a two-year assignment in Cambodia as the AFL-CIO's Country Representative. While in Cambodia she directed worker education and labor law training programs and conducted extensive research on the problems of women workers and on child labor. She is a social anthropologist, and received her Ph.D. from the University of Michigan. She spent three years in Indonesia, first as a State Department official and later as an independent researcher, and wrote her dissertation on Indonesia's labor movement. She has also lived and worked in China, Taiwan and India. Dr. Athreya has published extensively on the issues of corporate accountability and human rights in global supply chains, and has provided public commentary on CNN's Lou Dobbs Tonight, MSNBC, National Public Radio, and other major media outlets.

Toni Dembski-Brandl graduated from Carroll College in 1985 with a Bachelors of Arts in International Relations. In 1987, she received a Master of Art in Political Science from Marquette University. In 1996, Toni graduated from DePaul University College of Law and began her practice of law with one of the country's oldest international trade firms, Barnes, Richardson & Colburn. In 1997, Toni joined Target Corporation's Law Department. She advises clients in matters of Customs Law, International Trade, International Sale of Goods, International Finance and Social Compliance. Target Corporation is one of the largest importers of merchandise in the United States and has robust Customs Compliance and Social Compliance programs.

Charita L. Castro is Division Chief for the Program Operations and Research Team in U.S. Department of Labor's Office of Child Labor, Forced Labor, and Human Trafficking (OCFT). Dr. Castro leads a team of nine staff who are responsible for overseeing audits and evaluations of grants and contracts; funding extramural research and coordinating Congressionally-mandated reports such as the Department of Labor's *Annual Findings on the Worst Forms of Child Labor*; and managing the budget and reporting duties for the office, including submissions for the President's Management Agenda and the Government Performance and Results Act. Dr. Castro began her government career as a Presidential Management Fellow with the Children's Bureau of the U.S. Department of Health and Human Services. She holds a bachelor's degree in psychology from Tulane University, a master of social work from Washington University in St. Louis, and a doctorate from the George Washington University's School of Public Policy and Public Administration.

Bill Guyton is the president of the World Cocoa Foundation, private sector, non-profit organization focusing specifically on farmer outreach and environmental programs. Currently, Bill oversees regional public-private partnership programs that focus on cocoa sustainability. He also helps to coordinate an international research program with USDA and other partners. Bill assisted in forming the World Cocoa Foundation in July 2000, which now has over 60 chocolate companies and trade association members from North America, Europe, Latin America and Asia. Prior to joining the chocolate industry, Bill was Director of Business Development at the U.S. Grains Council where he helped to identify new market opportunities. Bill worked for more than 10 years in developing countries, advising and implementing agricultural and environmental programs for USAID, the World Bank, GTZ, OECD, Peace Corps and other development organizations. Bill holds a Master of Science Degree from Michigan State University in Agricultural Economics and a Bachelor of Science from Colorado State University.

Aurelie Hauchère has studied International Relations and has a Master in Development and Humanitarian Affairs. She has worked with several nongovernmental organizations and is now managing technical cooperation projects for the ILO Special Action Programme to Combat Forced Labour (www.ilo.org/forcedlabour), with a special focus on Latin America, France and French-speaking countries in Africa.

Thea Lee is policy director and chief international economist at the AFL-CIO, where she oversees research and strategies on domestic and international economic policy. Previously, she worked as an international trade economist at the Economic Policy Institute in Washington, D.C. and as an editor at Dollars & Sense magazine in Boston. She received a Bachelors degree from Smith College and a Masters degree in economics from the University of Michigan. Ms. Lee is co-author of *A Field Guide to the Global Economy,* published by the New Press. Her research projects include reports on the North American Free Trade Agreement, on the impact of international trade on U.S. wage inequality, and on the domestic steel and textile industries. She has appeared on numerous television and radio shows, including the News Hour with Jim Lehrer; CNN; Good Morning America; NPR's All Things Considered and Marketplace; and the PBS documentary, Commanding Heights. She has testified before several committees of the U.S. House of Representatives and the Senate on various economic policy topics. She serves on several advisory committees, including the State Department Advisory Committee on International Economic Policy and the Export-Import Bank Advisory Committee. She is also on the Board of Directors of the Worker Rights Consortium and the National Bureau of Economic Research.

Jeff Morgan is the director of Global Programs for Mars Incorporated, based at their global headquarters in McLean, Virginia, near Washington, DC. He has worked in the Food and Agriculture industry for nearly 33 years. He received his Bachelors Degree in Chemistry from Miami of Ohio and his Masters in Food Systems from The Ohio State University. He joined Mars Incorporated in 1979 where he has focused his attention on all aspects of the value chain for cocoa and chocolate. In that time he has worked on projects in nearly every major cocoa-growing region of Africa, Latin America and Asia – finding ways to improve cocoa farming practices and thereby benefit farmers, their families and their cocoa farming communities. Since 2005 Jeff has primarily worked to address questions related to Child and Adult labor practices in the cocoa sectors of Côte d'Ivoire and Ghana, representing not only the interests of Mars, Incorporated--but also representing the interests of the North American and European cocoa industry coalition-- the Global Issues Group (GIG). Within this effort, Jeff has worked in partnership with key stakeholders in Ghana and Côte d'Ivoire--being actively engaged on the ground and spending significant amounts of time in the West African Cocoa growing regions over the past 5 years. In addition to his effort on labor practices, Jeff is also working to implement a novel public / private partnership undertaken by Mars with a select group of Civil Society partners. The program, known as iMPACT (The Mars Partnership for African Cocoa Communities of Tomorrow) is focused on bringing needed services to cocoa growing communities in Ghana and Côte d'Ivoire, using a community empowerment model of entry, assessment and delivery.

Mark Neuman is counselor for International Trade & Global Strategies for Limited Brands (*www.LimitedBrands.com*) and MAST Industries (*www.MAST.com*). In that capacity, he serves as an international trade and country risk advisor to a leading specialty retailer and its global contract manufacturing and sourcing subsidiary with a dozen offices around the globe. Neuman is also responsible for leading an effort to develop links between women-owned organic cotton producers in West Africa and the Victoria's Secret Brand. Prior to his current position, Neuman served as a principal in the international investment banking and marketing firm, Global Linx Corporation. Previously, Neuman served on the Reagan White House Staff and also held a variety of positions on Capitol Hill and the Executive Branch -- including the Bureau of Export Administration, and on the Executive Staff of the US Bureau of the Census. Neuman remains active as an advisor to the U.S. Government, Congress and international organizations. He served as transition advisor to the new President of the World Bank, as an advisor to the Presidential transition team of the US President in 2001, and as chairman of the 2010 Census Advisory Committee, a joint appointment by the US Secretary of Commerce and Director of the Census. Neuman also served as staff liaison to the President's Advisory Committee for Presidential Trade Negotiations in the first term of the Bush Administration. Neuman currently serves on the Executive Committee of the National Cotton Board, having been appointed under both the Clinton and Bush Administrations. Previously, he co-chaired the presidentially mandated Census Monitoring Board, appointed by the U.S. Senate Majority Leader. Neuman is a graduate of the University of Illinois, Urbana with a degree in Economics and a native of Champaign-Urbana, Illinois.

Jorge Perez-Lopez is executive director of the Fair Labor Association (FLA), a multi-stakeholder organization that combines the efforts of industry, civil society organizations, and colleges and universities to protect workers' rights and improve working conditions worldwide by promoting adherence to international labor standards. He joined the FLA in 2005. Prior to joining the FLA, he spent 31 years in different positions within the U.S. Department of Labor, where he directed the Office of International Economic Affairs, Bureau of International Labor Affairs.

Rachel Phillips Rigby has been with the U.S. Department of Labor's Office of Child Labor, Forced Labor, and Human Trafficking (OCFT) for four years. She initially managed OCFT's grant portfolio in the South Asian subcontinent and conducted research on child labor and forced labor in that region, and now oversees DOL's research activities pursuant to the Trafficking Victims Protection Reauthorization Act of 2005, including the development of DOL's List of Goods made by forced labor and child labor. Ms. Rigby's background includes three years in the nonprofit sector with international microfinance and public health organizations. She holds an MBA from the Monterey Institute of International Studies, a B.A. in American Studies from Grinnell College, and has studied and volunteered abroad in Switzerland, France, England, and Bolivia.

Meg Roggensack is a lawyer specializing in international trade and human rights and is currently the Policy Director at Free the Slaves. During her private legal career at Hogan and Hartson, she headed the firm's Latin America Practice Group, guided the Community Service Department's expansion into public international law pro bono matters, and chaired coalitions of businesses, think tanks and nonprofit organizations. She has been recognized for her extensive pro bono work in the field of international human rights and justice reform. Ms. Roggensack serves on the boards of several human rights organizations, including the Due Process of Law Foundation and the Center for Refugee and Disaster Response at the Johns Hopkins Bloomberg School of Public Health. She is a past president of the Washington Foreign Law Society and a past Vice President of the Board of the Washington Office on Latin America. She is an adjunct professor at Georgetown University Law School, where she teaches "Human Rights at the Intersection of Trade and Corporate Responsibility."

Benjamin Smith is a chief technical advisor (CTA) for the ILO's International Programme on the Elimination of Child Labour (IPEC), and serves as IPEC Focal Point for Corporate Social Responsibility and Child Labour. He heads IPEC's Project TACKLE, which supports efforts to eliminate child labour through education in 11 countries in Africa, the Caribbean and the Pacific and is IPEC's single largest technical cooperation project. He began his career with IPEC as a Programme Officer in San Jose, Costa Rica in 2000. From 2002-2006, he served as the CTA for the project of support to the Time Bound Programme in El Salvador, one of three initial countries to launch this integrated modality to tackle child labour. Prior to joining the ILO, he was an International Programme Analyst at the U.S. Department of Labor and a Presidential Management Fellow. He holds a B.A. in International Studies from American University and an M.A. in International Relations from Johns Hopkins University.

Anna Walker is senior manager of Worldwide Government Affairs and Public Policy at Levi Strauss & Co. She is responsible for developing, planning and delivering government and public policy initiatives on environmental sustainability, worker rights, human rights and other business issues to promote LS&CO.'s business success. Prior to joining Levi Strauss & Co., Ms. Walker served for four years as Manager, Labor Affairs and Corporate Responsibility with the U.S. Council for International Business in New York City, where she advised member companies on international labor policy issues and international developments in the corporate citizenship arena. Ms. Walker also represented U.S. business at the International Labor Conference and the Governing Body of the International Labor Organization. Ms. Walker holds an M.A. in International Relations and Economics from Johns Hopkins University Paul H. Nitze School of Advanced International Studies and B.A.s in International Relations and Spanish from the University of California, Davis.

Vicki Walker is with Winrock International in the Empowerment and Civic Engagement unit. She was the Director of the Community-based Innovations to Reduce Child Labor through Education (CIRCLE) project funded by the U.S. Department of Labor (USDOL) which documented and published a manual *Best Practices in Eliminating Child Labor through Education* drawn from the global CIRCLE projects. She leads the innovative program Child Labor Alternatives through Sustainable Systems in Education (CLASSE) in Côte d'Ivoire. CLASSE has worked in Mali and Côte d'Ivoire to address the push and pull factors in child labor in cocoa to address prevention of Child labor through educational alternatives including vocational education in agriculture that contribute to training a new generation of leaders and agricultural entrepreneurs. Ms. Walker leads the Winrock component of the ECHOES project, a public private partnership with the World Cocoa Foundation and members and the U.S. Agency for International Development (USAID) that has expanded the agricultural vocational educational principles for livelihoods in Côte d'Ivoire and Ghana. She is currently the home office coordinator and advisor to the Tanzania Education Alternatives for Children funded by USDOL. Ms. Walker has BA from the University of Maryland and a MA from the University of New Hampshire, both in Political Science and served in the Peace Corps in Senegal. Ms. Walker's areas of focus are leadership, gender analysis, agriculture, prevention and elimination of child labor, promotion of women and children's education, and improving economic status of rural communities and youth.

Appendix D
Workshop Participants

First name	Last name	Affiliation
Sheela	Ahluwalia	World Bank
Bama	Athreya	International Labor Rights Forum
Kevin	Bales	Free the Slaves
Susan	Berkowitz	Westat, Inc.
Sherilyn	Brodersen	Cadbury
Brian	Campbell	International Labor Rights Forum
Charita	Castro	U.S. Department of Labor
Donna	Chung	Sandler, Travis, and Rosenberg
Marina	Colby	International Labor Organization
Tu	Dang	U.S. Department of State
Toni	Dembski	Target
Pamela	Dieguez	U.S. Department of Homeland Security
Kimberly	Elliott	Center for Global Development
John	Francis	
Alison Kiehl	Friedman	ASSET
Elaine	Fultz	Independent Consultant
Adam	Greene	U.S. Council for International Business
Bill	Guyton	World Cocoa Foundation
Aurélie	Hauchère	International Labor Organization
Patricia	Jurewicz	As You Sow
Thea	Lee	AFL-CIO
Jessica	Leslie	Free the Slaves
Beryl	Levinger	Monterey Institute of International Studies
Shawn	MacDonald	Verite
Reid	Maki	Child Labor Coalition
Theodore	Moran	Georgetown University
Jeff	Morgan	Mars, Inc.
Mark	Neuman	
Tim	Newman	International Labor Rights Forum
Anthony	Ng	Macy's
Amy	O'Neill Richard	U.S. Department of State
Monique	Oxender	Ford Motor Company
John	Pacheco	C & M International
Jorge	Perez-Lopez	Fair Labor Association
Desta	Raines	Jones Apparel Group
Rachel	Rigby	U.S. Department of Labor
Meg	Roggensack	Free the Slaves
Ruth	Rosenbaum	CREA

APPENDIX D

Brandie	Sasser	U.S. Department of Labor
Mira	Shastry	C & M International
Benjamin	Smith	International Labor Organization
Leyla	Strotkamp	U.S. Department of Labor
Kathy	Ting	U.S. Department of Agriculture
Maurizia	Tovo	World Bank
Dan	Viederman	Verite
Anna	Walker	Levi Strauss
Vicki	Walker	Winrock
Laurie	Weeks	U.S. Department of Homeland Security
Kevin	Willcutts	U.S. Department of Labor

Appendix E
Definitions of Child and Forced Labor
by John Sislin

Any attempt to assess the efficacy of business practices to reduce the use of child or forced labor must be grounded in an understanding of the nature of these types of labor, in particular their definitions, scope, and causes. At the heart of the notion of what is and is not child and forced labor, lies the core conventions established by the International Labor Organization (ILO).[1] The ILO is an international organization, founded in 1919 and currently an agency within the United Nations, is comprised of 182 member states as of April 2009.[2] In 1995, the ILO identified four rights as "fundamental to the rights of human beings at work. These rights are freedom of association and the right to organize and bargain collectively, the abolition of forced labor, equal remuneration and nondiscrimination in employment, and the elimination of child labor.

One mechanism of the ILO consists of establishing Conventions, or international treaties, subject to ratification by ILO member States. In May 1995, the Director-General of the ILO launched a campaign to achieve universal ratification of the then seven core Conventions which were designed to support these four rights. The Conventions are: Nos. 87 and 98 on freedom of association and the right to organize and bargain collectively, Nos. 29 and 105 on forced labor, Nos. 100 and 111 on equality, and No. 138 on child labor.

"In 1998 the ILO adopted a Declaration on Fundamental Principles and Rights at Work, committing ILO member nations to realize and achieve at the policy level the four basic rights as an obligation inherent in ILO membership, regardless of whether or not they have ratified conventions corresponding to those rights."[3] Finally, in 1999, the ILO adopted Convention No. 182, on the worst forms of child labor. The four Conventions on child and forced labor are the starting point for defining what this labor is.

According to Swepson, child labor had not been developed "as a core human rights subject, or even as a subject of great concern" until the last two decades of the 20th century.[4] Child labor falls into three categories: some is considered good for a child's development (e.g., apprenticeship or school programs), some is acceptable, and some is

---

[1] This report uses American spelling, hence "labor" and not "labour." Those interested in further researching these topics should note that search engines and other databases can return different results depending on the spelling of the terms.

[2] For a list of Member States, see: ILO, "Alphabetical list of ILO member countries" available at: *http://www.ilo.org/public/english/standards/relm/country.htm*. Accessed May 17, 2009.

[3] National Research Council, Monitoring International Labor Standards: Techniques and Sources of Information. Washington, DC: National Academies Press, 2004:17-18. For the text of the Declaration, see ILO, "The Declaration," available at: *http://www.ilo.org/declaration/thedeclaration/textdeclaration/lang--en/index.htm* Accessed May 17, 2009.

[4] Lee Swepston. "The Contribution of the ILO Declaration on Fundamental Principles and Rights at Work to the Elimination of Child Labour." Child Labour in a Globalized World, Giuseppe Nesi, Luca Nogler, and Marco Pertile, eds. Surry, UK: Ashgate, 2008, p. 68. Basu and Tzannatos disagree, saying societies have been trying to root out child labor at least as far back as 1802. Kaushik Basu and Zafiris Tzannatos, "The Global Child Labor Problem: What Do We Know and What Can We Do?" The World Bank Economic Review, 17:2, 2003, pp. 147-173.

unacceptable. ILO Conventions Nos. 138 and 182 delimit child work and in particular to highlight what forms should be eliminated.

Convention No. 138 was adopted in 1973 and establishes minimum ages for work by children. Members who ratify this Convention are directed to specify a minimum age for admission to employment or work. The Convention text further states that the minimum age specified "shall not be less than the age of completion of compulsory schooling and, in any case, shall not be less than 15 years." Furthermore, "The minimum age for admission to any type of employment or work which by its nature or the circumstances in which it is carried out is likely to jeopardize the health, safety or morals of young persons shall not be less than 18 years." The specification of hazardous types of employment or work is determined by national sources. Thus, multiple minimum ages were specified.

Two exceptions identify work which is permissible. First, the "Convention does not apply to work done by children and young persons in schools for general, vocational or technical education or in other training institutions, or to work done by persons at least 14 years of age in undertakings, where such work is carried out in accordance with conditions prescribed by the competent authority, after consultation with the organizations of employers and workers concerned, where such exist, and is an integral part of--(a) a course of education or training for which a school or training institution is primarily responsible; (b) a program of training mainly or entirely in an undertaking, which program has been approved by the competent authority; or (c) a program of guidance or orientation designed to facilitate the choice of an occupation or of a line of training."

Second, "national laws or regulations may permit the employment or work of persons 13 to 15 years of age on light work which is--(a) not likely to be harmful to their health or development; and (b) not such as to prejudice their attendance at school, their participation in vocational orientation or training programs approved by the competent authority or their capacity to benefit from the instruction received." Note the lack of definition of what "light work" *is*, and such key notions as defining when it is harmful to health or development or prejudicial, for instance.[5]

Convention No. 182 identifies the worst forms of child labor that should be abolished. These are: (a) all forms of slavery or practices similar to slavery, such as the sale and trafficking of children, debt bondage and serfdom and forced or compulsory labor, including forced or compulsory recruitment of children for use in armed conflict; (b) the use, procuring or offering of a child for prostitution, for the production of pornography or for pornographic performances; (c) the use, procuring or offering of a child for illicit activities, in particular for the production and trafficking of drugs as defined in the relevant international treaties; (d) work which, by its nature or the circumstances in which it is carried out, is likely to harm the health, safety or morals of children. Clause (d) remains the most ambiguous and is determined within a state's national context.

Forced labor—particularly adult forced labor—is a very different phenomenon than child labor and requires a different policy response. Children engaged in prohibited activity have ideally to be withdrawn from the labor market and provided acceptable

---

[5] For a discussion of light work, see: Augendra Bhukuth, "Defining child labour: a controversial debate" Development in Practice, Volume 18, Number 3, June 2008, pp. 385-394.

alternatives such as education. For adults, the challenge is to keep them in the labor market, possibly even in the same jobs, while tackling the coercive elements in the recruitment and employment relationship that locks them into bondage and exploitation

According to the ILO, historically, action against forced labor focused on slavery; and the first international agreement focused on slavery was the Declaration Relative to the Universal Abolition of the Slave Trade, adopted in 1815 by the Congress of Vienna.[6] "Forced labour issues as such became the subject of systematic study and standard setting at the international level only after the First World War, following the work of the League of Nations regarding mandated territories and of the adoption of the 1926 Slavery Convention."[7] Forced labor is defined by the ILO in Convention No. 29, which was adopted in 1930. According to the treaty language, "forced or compulsory labor shall mean all work or service which is exacted from any person under the menace of any penalty and for which the said person has not offered himself voluntarily." As Belser et al note: "Embedded in the international definition of forced labour as formulated in ILO Convention No. 29 are two essential criteria: 'menace of penalty' and 'involuntariness'. Accordingly, forced labour occurs when people are being subjected to psychological or physical coercion (the menace or the imposition of a penalty) to perform some work that they would otherwise not have accepted to perform at the prevailing conditions (the involuntariness). The use of deception or fraud, and the retention of identity documents in order to achieve the consent of workers, are illegitimate and can lead to forced labour."[8]

Exceptions to the above definition include: "(a) any work or service exacted in virtue of compulsory military service laws for work of a purely military character; (b) any work or service which forms part of the normal civic obligations of the citizens of a fully self-governing country; (c) any work or service exacted from any person as a consequence of a conviction in a court of law, provided that the said work or service is carried out under the supervision and control of a public authority and that the said person is not hired to or placed at the disposal of private individuals, companies or associations; (d) any work or service exacted in cases of emergency, that is to say, in the event of war or of a calamity or threatened calamity, such as fire, flood, famine, earthquake, violent epidemic or epizootic diseases, invasion by animal, insect or vegetable pests, and in general any circumstance that would endanger the existence or the well-being of the whole or part of the population; (e) minor communal services of a kind which, being performed by the members of the community in the direct interest of the said community, can therefore be considered as normal civic obligations incumbent upon the members of the community, provided that the members of the community or their

---

[6] Declaration Relative to the Universal Abolition of the Slave Trade, 8 February 1815, Consolidated Treaty Series, Vol. 63, No. 473. Cited in *Eradication of forced labour* - General Survey concerning the Forced Labour Convention, 1930 (No. 29), and the Abolition of Forced Labour Convention, 1957 (No. 105). Report of the Committee of Experts on the Application of Conventions and Recommendations, Report III (Part Ib), ILC, 96th Session, Geneva, 2007.
[7] *Eradication of forced labour* - General Survey concerning the Forced Labour Convention, 1930 (No. 29), and the Abolition of Forced Labour Convention, 1957 (No. 105). Report of the Committee of Experts on the Application of Conventions and Recommendations, Report III (Part Ib), ILC, 96th Session, Geneva, 2007.
[8] Patrick Belser, Michaëlle de Cock, and Farhad Mehran. ILO Minimum Estimate of Forced Labour in the World. Geneva: ILO, 2005:6-7.

- Economic participation by children;
- Full-time work performed by children;
- Work that is harmful to children;
- Work that interferes with schooling;
- All remunerated work;
- Wage employment;
- Work that exploits children;
- Work that violates national child labour laws;
- Work that violates international standards.

This sampling of recent definitions does not describe a single phenomenon. The different definitions imply quite dissimilar notions about just what is problematic about 'child labour', and they of course lead to divergent policies and activities for addressing the issue. For example, a strategy to halt all work by children would not necessarily resemble one seeking to discourage only wage employment, or only work that is harmful to children.[16]

Also, operationalizing concepts can be tricky. (This is crucial in trying to measure the incidence of child or forced labor.) In the realm of child labor, operationalizations deal with terms such as child work, child labor, and child economic activity. According to the ILO:

Economic activity" is a broad concept that encompasses most productive activities undertaken by children, whether for the market or not, paid or unpaid, for a few hours or full time, on a casual or regular basis, legal or illegal; it excludes chores undertaken in the child's own household and schooling. To be counted as economically active, a child must have worked for at least one hour on any day during a seven-day reference period. "Economically active children" is a statistical rather than a legal notion. "Child labour" is a narrower concept than "economically active children", excluding all those children aged 12 years and older who are working only a few hours a week in permitted light work and those aged 15 years and above whose work is not classified as "hazardous". The concept of "child labour" is based on the ILO Minimum Age Convention, 1973 (No. 138), which represents the most comprehensive and authoritative international definition of minimum age for admission to employment or work, implying "economic activity". "Hazardous work" by children is any activity or occupation that, by its nature or type, has or leads to adverse effects on the child's safety, health (physical or mental) and moral development.[17]

---

[16] William E. Myers, "Appreciating Diverse Approaches To Child Labor." A presentation during the symposium "Child Labor & the Globalizing Economy: Lessons from Asia/Pacific Countries," Stanford University, California, February 7 – 9, 2001. Available at: http://www.childlabor.org/symposium/myers.htm.

[17] International Labor Organization, *The End of Child Labour: Within reach*, Global Report under the follow-up to the ILO Declaration on Fundamental Principles and Rights at Work, International Labour Conference, 95th Session 2006, Report I (B). Geneva: ILO, 2006:6. For a further discussion of what

Ritualo et al. identify a variety of large scale efforts to collect data on child labor, for example by the ILO, the World Bank, and by national countries. While these data are indeed useful to get a big picture view of child labor, a question is the degree to which these approaches are helpful vis-à-vis business practices to eliminate child or forced labor in specific situations.

---

economic activity is, see: Amy Ritualo, Charita Castro and Sarah Gormly, "Measuring Child Labor: Implications for Policy and Program Design", Comparative Labor Law & Policy Journal v. 24 no2 (Winter 2003) p. 401-434.

Appendix F
Illustrative Examples of Business Practices
by Kara Murphy and John Sislin

Introduction

The following are illustrative examples of business practices and partnerships that aim to reduce the use of forced or child labor in the production of goods. The practices included in this appendix have not been evaluated as part of the process for preparing this report so their inclusion does not convey any representation of their quality or effectiveness or the accuracy of company statements or statements from non-governmental organizations about company practices included in the text.

The illustrative examples have been adapted from reports by international organizations (such as the International Labour Organization, World Bank, and UNICEF), non-governmental organizations and labor associations, and from the companies themselves (social responsibility reports and websites) in order to illustrate the five categories of business practices described below. "Practice" is understood to include "policy, planning and research activities, legislation, programs and projects, as well as "on-the-ground" delivery of programs."[1]

Definitions

1) Policy-making, code of conduct, guidelines

A company's effort to reduce child or forced labor often begins with its code of conduct. The code of conduct is a formal statement of the policies that a company requires its employees—both on- and off-site—to adhere to. In other words, the code of conduct applies to the entire supply chain. For example, a code of conduct may include a minimum age requirement or a provision against prison labor. According to the International Labour Organization, "The concept 'corporate code of conduct' refers to companies' policy statements that define ethical standards for their conduct. There is a great variance in the ways these statements are drafted."[2] According to the World Bank, codes came about in the 1970s in response to declarations of the OECD and the ILO encouraging multinational companies to be more socially responsible. These codes set guidelines for a broad range of business practices. In the 1990s, corporations began to include a focus on forced and child labor.[3]

Policy-making refers to the process of developing policies to combat child or forced labor. Policies may reflect or comply with international labor standards, as well as national and local labor laws in the country. Companies also sign on to collective policy statements by a group of companies or employers' organizations in order to strengthen their own practices.

---

[1] *Combating Child Labour: Sample Good Practices Guidelines*. Understanding Children's Work (UCW). Workshop Report, October 2003.
[2] ILO, *http://actrav.itcilo.org/actrav-english/telearn/global/ilo/code/main.htm*.
[3] The World Bank, Labor Markets Codes of Conduct (online).

Guidelines can refer to a variety of business practices. Companies may follow the guidelines of multi-stakeholder initiatives or industrial associations that they join. Guidelines can also be the recommendations that a company makes to its contractors and sub-contractors. Additionally, a company may develop guidelines with the goal of sharing them as "good practices" with others in the same industrial sector.

2) Data collection, research, risk-assessment, understanding and communicating information

In order to implement a preventative practice effectively, a company should understand the on-the-ground realities where production is taking place. Each country, community, and production facility is unique. For this reason, research and data-collection are crucial to a company's understanding of a potentially harmful situation. Companies that seek better understanding of local situations may conduct in-house research, perform root-cause analysis, administer surveys, or conduct interviews.

A risk-assessment evaluates the risk that forced or child labor will occur. When entering a new market, setting up a production facility, or contracting with a new vendor, a company may seek to determine which workers are the most vulnerable. A risk-assessment is often carried out in conjunction with NGOs, trade unions and governments. The ILO identifies characteristics of vulnerable workers: ethnic or minority background, poverty, age, gender (female), irregular migrant status, lack of skills, illiteracy, and employment in the informal sector.[4] A company may also need to assess the risk that employers or factory owners will not comply with its policies. Businesses should be aware of the risk to their own welfare if they do not take action. According to the ILO, "To be successful, companies must manage risk in an environment where risk is not static and can emerge through the actions of the company itself, its suppliers and other actors. Allegations of forced labour and trafficking present legal risks as well as serious threats to brand and company reputation."[5]

Finally, communication involves raising awareness of labor rights and conditions among workers and their communities. In terms of child labor, companies may face the hurdle of convincing employers and families that child labor has serious harmful effects. In that case, one priority may be to transform the "culture of work" in that community.[6] Other communication strategies include lobbying local or national governments, and communicating practices, both successful and not, to the wider global community.

3) Changing business practices

Companies may change their business organization or operations to address child or forced labor. A variety of reasons can prompt a company to make systematic changes. These reasons may include pressure from the international community or labor rights organizations, negative publicity, or discovery of child labor where it was not known before. In addition, companies may encourage each other if they believe they have taken actions that are effective and can be replicated in other companies in the same or different sectors.

---

[4] *Combating Forced Labour: A Handbook for Employers and Business*, International Labour Organization, 2008.
[5] ILO Handbook for Employers, 2008.
[6] Good Practice Note, IFC.

APPENDIX F

With multiple stages of production and factories across the globe, companies look for ways to exert more control over their production. Changes to supply chain operations can include consolidating production, training business partners, creating incentives, and reducing hazardous risks in the workplace. In some cases, a company may even elect to boycott a certain product. Internal operations can be modified as well. A company can change its organizational structure by adding a new department of social responsibility, for example. Company annual reports may also reflect this by devoting more space to social responsibility reporting. As companies try to become more transparent, they may also publish data and findings from audits.

4) Monitoring, compliance, enforcement and certification

Policies prohibiting child and forced labor are enhanced through enforcement efforts. Without enforcement, it is unlikely that a company's policy would be translated into reality. Enforcement practices include monitoring, auditing, compliance, verification, and certification. According to the ILO, "an effective social audit can make an important contribution to the identification, prevention and eradication of forced labour".[7]

Monitoring may be internal or external. The goal of external monitoring is to bring in an objective, outside person or group to see that contractors are complying with company standards. Companies may contract with the Fair Labor Association or Verite accredited monitors who follow the workplace codes of conduct. Companies also seek guidance from NGOs and associations such as the Ethical Trading Initiative to implement their enforcement practices. Partnerships with independent certification programs are becoming more common, but have produced mixed results in terms of effectiveness.[8]

5) Remediation

Remediation refers to a variety of practices. It may be an action a company takes to remedy non-conformance of a contractor or factory. In relation to child labor, it refers to "all support and actions necessary to ensure the safety, health, education, and development of children who have been subjected to child labour...and have been subsequently dismissed."[9]

Remediation is a complex and sensitive process. As UNICEF states:
Early calls for the elimination of child labour sometimes resulted in large numbers of young workers being summarily dismissed with no recourse to an alternative income. Since then corporate strategies to deal with child labour have evolved from a "cut and run" response to more responsible engagement with the community where child labour is present.[10]

When child or forced labor is found, a company faces the choice of terminating a contract with the offending supplier, or taking action to improve the working conditions.

Remediation has received new focus because of criticism of companies that terminate contracts without taking into consideration the potential negative effects on workers. Through remediation, companies can try to address the root causes of the

---

[7] ILO Handbook for Employers, 2008
[8] Good Practice Note, IFC
[9] Social Accountability 8000:2008, Social Accountability International
[10] UNICEF UK's Child Labour Resource Guide, Executive Summary

harmful working conditions in a factory.[11] In seeking remedial action, the ILO encourages businesses to work with NGOs and labor unions.[12]

When children or forced laborers are withdrawn from work, the next step is to provide them with a transition. The transition may be back to their family (if they were physically separated) and/or into a school environment. A variety of remedial actions can be taken. A business can implement educational programs, pay for transfer from work to school, tend to the person's physical and mental health, fund programs that provide services to families and children, and address conditions that prompted child or forced labor in the first place. Companies are also encouraged not to ignore children of legal working age, and to ensure that they too have access to education.[13] Whatever approach or approaches a company chooses, a key element is whether or not the practice is sustainable for the community and for the workers.[14]

Illustrative Examples

1) Policy-making, code of conduct, guidelines

Levi Strauss & Co.

In 1991, Levi Strauss became the first multi-national apparel company to establish a code of conduct. With Levi Strauss and others leading the way, the apparel industry made significant progress in the 1990s in terms of applying core labor standards to business practices.[15] Levi's code, Global Sourcing and Operating Guidelines (GSOG), is made up of two parts: Country Assessments and Business Partner Terms of Engagement. The code is based on the Universal Declaration of Human Rights and many of the International Labor Organization's (ILO) Core Conventions.[16]

With approximately 600 external suppliers in more than 50 countries, Levis acknowledges the gap that can exist between the Code of Conduct and the actual practice of suppliers. Since 1991, Levi's has modified its Code to incorporate lessons it learned from experience as well as feedback from NGOs.

Freedom of Association is one area in which Levi's has made changes. The company Code reads:

> Factories must respect employee rights to freedom of association; they must not impose any punitive actions against workers in supporting union such as threatening, fining, suspending or firing workers exercising those rights. Any action that suppresses freedom of association is prohibited, and may be an act deemed illegal in some countries' labor codes.[17]

Despite this policy, Levi's observed resistance from some managers in countries such as Bangladesh, Cambodia, the Dominican Republic, Haiti and Mexico. To

---

[11] The World Bank, Labor Markets Codes of Conduct (online)
[12] ILO Handbook for Employers, 2008
[13] Good Practice Note, IFC.
[14] Social Accountability International.
[15] Monitoring International Labor Standards: Summary of Domestic Forums, National Research Council, 2003.
[16] Levi Strauss Web site: *http://www.levistrauss.com/Citizenship/ProductSourcing.aspx* .
[17] Levi Strauss & Co., Terms of Engagement Guidebook, 2007.

APPENDIX F

strengthen freedom of association, the company drew from the Ethical Trading Initiative's guidelines as well as advice from labor and human rights organizations. The new Terms of Engagement included language on lawful "parallel means" for independent free association and bargaining:

> Where the right to freedom of association and collective bargaining is restricted under law, the supplier should not hinder the development of lawful parallel means for independent free association and bargaining.

To see that these changes are applied, Levis partnered with Cornell University's School of Industrial and Labor Relations. The school provides training to Levi's managers. Levi's also developed locally-tailored freedom of association training for external monitors and suppliers.

Starbucks Coffee Company

Starbucks developed the Cocoa Practices to "provide a set of comprehensive sustainability guidelines for cocoa."[18] With these guidelines, Starbucks hopes to promote socially responsible cocoa production throughout its supply chain as well as the supply chains of other companies. The Cocoa Practices include minimum participation requirements for producers as well as a highly-detailed evaluation checklist. Starbucks' approach aims to "enhance buying companies' ability to work with a diversity of suppliers to foster deeper sustainability throughout the cocoa supply chain from farm to factory to final product."

The guidelines resulted from discussions with a variety of stakeholders including Theo Chocolate staff, C.A.F.E. (Coffee and Farmer Equity), environmental and labor organizations, and cocoa suppliers. In 2007, Starbucks initiated a two-year pilot program to test the standards. In addition, the International Labor Rights Forum has approved of the Starbucks Cocoa Practices, which it says are in line with ILO standards on child and forced labor.[19]

The Starbucks guidelines set minimum requirements for participation in the program. The main tool of the program is the Cocoa Practices Scorecard, which has two tiers of required indicators: Zero Tolerance Indicators and Criteria Requirement Indicators. The indicators include product quality, economic accountability, internal control systems, organizational stability, and social responsibility. The indicators are scored as Comply, Not Comply or Not Applicable.

An example of a Cocoa Practices guideline is hiring practices and employment policies in relation to child/forced labor:

> *Objective:* No direct contracting of any persons under the age of 14 (ILO Convention 138). (We prefer that our suppliers hire no one under the age of 15). If local regulations stipulate compulsory education up to an age greater than 15, those regulations will apply during school hours. Children of producer families can participate in activities which are not harmful to their health and do not conflict with local educational calendar and schedules. The use of any forced, involuntary or trafficked labor, either directly or indirectly, by our suppliers,

---

[18] Starbucks Coffee Company, Cocoa Practices: Evaluation Guidelines (April 1, 2008) Version 1.1.2.
[19] Chocolate Company Scorecard: The Sweet and the Bitter, February 14, 2009 *http://www.laborrights.org/files/ChocolateScorecard09.pdf.*

contractors or subcontractors, producer organizations, etc. will not be tolerated (ILO Convention 182 and 29).

Wal-Mart

Wal-Mart is the world's largest retailer and corporation. In 1992, the company established Standards for Suppliers. Wal-Mart's Code says that neither forced/prison labor nor child labor will be tolerated. On child labor, "Wal-Mart will not accept products from suppliers or subcontractors who use child labor. No person shall be employed at an age younger than the legal minimum age for working in any specific jurisdiction. In no event shall suppliers or their subcontractors employ workers less than 14 years of age."

In 2005, the International Labor Rights Forum (ILRF) filed a class-action lawsuit against Wal-Mart Stores. In the lawsuit, apparel workers in Bangladesh, China, Indonesia and elsewhere said that Wal-Mart did not enforce its code of conduct for overseas contractors: "Based on its vast economic power, Wal-Mart, based on its code of conduct, can and does control the working conditions of its supplier factories .... It could use its power and position to prevent its producers from profiting from the inhumane treatment of plaintiffs."[20]

Since 2005, Wal-Mart has taken steps to improve communication of its code of conduct to suppliers. The company engages its suppliers in training sessions on Wal-Mart's Ethical Standards Program as well as industry-wide best practices. In 2007, more than 10,100 suppliers and factory management attended training sessions.[21] However, ILRF maintains that Wal-Mart continues to cover up labor violations, including forced labor and the denial of the right to form independent unions.[22]

The Dress Barn

The Dress Barn is a leading specialty apparel retailer that has developed a code that applies to vendors who sell directly or indirectly to them. While it is not a participant in the United Nations Global Compact, the Dress Barn Global Human Rights Policy outlines its policy on child and forced labor.[23] Before doing business with a factory, the Dress Barn requires the factory to complete a questionnaire to ensure that the factory can abide by its code of conduct.

The Dress Barn's Vendor Code of Conduct reads:[24]
Vendors must operate in full compliance with the laws of their respective countries and with all other applicable laws, rules and regulations.
A. Child Labor:
Vendors shall only employ workers who meet the applicable minimum legal age requirements by local law or who are at least 15 years of age or who are older than the age for completing compulsory education in the country of manufacture, whichever is greater.

---

[20] "Suit Says Wal-Mart Is Lax on Labor Abuses Overseas," New York Times, September 14, 2005.
[21] Ethical Standards Program Fact Sheet, Wal-Mart.
[22] International Labor Rights Forum, Wal-Mart Campaign, *http://www.laborrights.org*.
[23] *http://www.business-humanrights.org/Categories/Individualcompanies/D/DressBarn*.
[24] The Dress Barn Vendor Code of Conduct, *http://www.dressbarn.com/pdf/VendorCodeOfConduct.pdf*.

Vendors must also comply with all other applicable child labor laws, including those related to hiring, wages, hours worked, overtime and working conditions. Vendors shall maintain official documentation for every worker that verifies the worker's date of birth. In those countries where official documents are not available to confirm exact date of birth, the vendor shall confirm the age using appropriate and reliable assessment methods. Vendors are encouraged to develop lawful workplace apprenticeship programs for the educational benefit of their workers.

B. Forced Labor:

Vendors shall not use any prison, slave, indentured, bonded or forced labor in the production of any product, including labor required as a means of political coercion or punishment for expression of political views in its production or manufacture. Vendors who recruit foreign contract workers must pay agency recruitment commissions and must not require any worker to remain in employment for any period of time against his or her will.

2) Data collection, research, risk-assessment, understanding and communicating information

Patagonia, Inc.

Patagonia is a privately owned designer, marketer, and retail seller of apparel and climbing equipment. It does not own any of the over 100 manufacturing facilities that make its products. "The key area where Patagonia can influence its supply chain is by determining who we are doing business with. Recently we have adopted a fourfold approach to all of our sourcing and factory decisions – one that gives equal weight to business requirements, quality assurance, social responsibility and our environmental footprint."[25]

Because Patagonia outsources its entire production, a first challenge is to decide which plants to use. Patagonia avoids "transitional" plants—those that have had problems, even minor, in the past. The company usually will source only from a factory that is currently being monitored.[26] In addition, Patagonia considers the factors that increase the probability of code violations. These factors include the country the factory operates in, the products they make and the age of the factory.[27] When analyzing a plant, Patagonia also calculates the reputational risk involved in sourcing from a plant that may be outstanding in all but one area. In contrast to larger companies, Patagonia's small production runs do not give it the same degree of leverage with factories as that of larger apparel and retail companies. With this limited leverage, the company cannot force a factory's management to address international labor standards violations because the cost of losing Patagonia's business is insignificant for any large manufacturing facility.

---

[25] Patagonia and Corporate Social Responsibility, November 2007
[26] Monitoring International Labor Standards: Summary of Domestic Forums, National Research Council, 2003.
[27] Patagonia and Corporate Social Responsibility, November 2007

Ford Motor Company

Ford works to understand the conditions in the communities where it operates. The company believes that this understanding is essential to a good human rights policy. As UNICEF recommends, "Businesses pursuing this goal have to take the background conditions in the areas in which they operate as a given, and work within these constraints".[28]

In order to understand the conditions, the company engages members of the local communities during site visits, interviews facility management, and works closely with a variety of stakeholders. Ford found that an interactive, collaborative approach to working with suppliers helped the company better understand the working conditions. At the same time, Ford is aware of the systemic factors underlying child and forced labor. The company promotes root cause analysis as key to understanding the system that allows or contributes child and forced labor:

> Reduction of these labor abuses is not easy to measure and should not be a single point in time measurement. Also included should be a rate of recurrence over x amount of time to account for cyclical behaviors. Take the example of forced labor in Brazil. One out of every 10 temporary work opportunities may result in a forced labor situation but due to the need for employment and income, lack of education as well as lack of interest in more permanent employment, laborers are willing to take that chance. And this cycle plays out over a period of months to years.[29]

To communicate its findings, Ford reports to human rights groups and then publishes its findings on its website for the public. In addition, Ford asks neutral third parties to review its assessment process. Some of the third-party reviewers have first-hand experience at Ford's plants. Ford stresses that the consensus from third-party review affirms that Ford's process *"is robust and has integrity"*.[30]

Merck & Co, Inc.

According to its website, "Merck & Co., Inc., is a global research-driven pharmaceutical company dedicated to putting patients first. Established in 1891, Merck discovers, develops, manufactures and markets vaccines and medicines to address unmet medical needs." Merck is committed to human rights, and has not found major violations in its operations. The company maintains that it is not at significant risk of forced or compulsory labor, incidents of child labor, or violations of the right to exercise freedom of association and collective bargaining. Still, Merck is taking steps to research, understand, and foster communication about child and forced labor. Merck collaborates with other pharmaceutical companies in two key initiatives.

First, Merck works with the Danish Institute for Human Rights to define its health-related human rights obligations. The Danish Institute is assisting Merck and

---

[28] UNICEF UK's Child Labour Resource Guide, Executive Summary
[29] Monique Oxender, Global Manager, Supply Chain Sustainability Ford Motor Company, email correspondence
[30] *For A More Sustainable Future,* Ford Motor Company Sustainability Report 2006/7

other pharmaceutical companies to explore the possibility of a sector-specific human rights assessment tool.[31]

In addition, Merck is a founding member of Pharmaceutical Supply Chain Initiative, a group of major pharmaceutical companies that aim to "help ensure that working conditions in the pharmaceutical supply chain are safe and that workers are treated with respect and dignity."[32] In addition to better conditions for workers, PSCI promotes economic development and a cleaner environment for local communities. PSCI explicitly states, "Suppliers shall not use child labor. The employment of young workers below the age of 18 shall only occur in non hazardous work and when young workers are above a country's legal age for employment or the age established for completing compulsory education".

3) Changing Business Practices

Reebok

In 1996, the International Labor Rights Forum (ILRF) led a campaign to shed light on the thousands of children ages 5 and 14 who were stitching soccer balls in Pakistan. Children were observed working as many as 10 to 11 hours per day.[33] Reebok responded by setting a goal to eliminate child labor in its soccer ball supply chain. Because children were doing the stitching at home, Reebok required its supplier to move soccer ball production to a central facility that could be regularly monitored: "We devised a unique plan to end this practice by consolidating all ball stitching in a newly constructed factory dedicated to Reebok ball production that guaranteed manufacturing without child labor. This initiative, another first of its kind in the industry, included a program to raise funds for the educational and vocational needs of children in Pakistan through the sale of Reebok soccer balls."[34]

With better oversight of the manufacturing facility, Reebok's "No Child Labor" program was able to implement labeling certification. Soccer balls produced without child labor bear the label "Guaranteed: Manufactured Without Child Labor." Reebok's program earned it the 1996 Business Ethics Award for Corporate Social Responsibility. Reebok continues to require full disclosure of its business partners involved in production.

Monsanto Corporation

Monsanto Corporation is an agricultural company that is involved in efforts to address child labor within the cotton seed industry. With the urging of groups such as the International Labor Rights Forum, Monsanto undertook a global risk assessment and child labor monitoring program in India. The Monsanto Human Rights Policy, established in 2006, is based on Universal Declaration of Human Rights and ILO Declaration on Fundamental Principles and Rights at Work. The Employee Guidebook

---

[31] Merck & Co, Inc., Corporate Social Responsibility 2006-2007 Report, *http://www.merck.com/corporate-responsibility/docs/cr2006-2007.pdf*
[32] *http://www.pharmaceuticalsupplychain.org/*
[33] International Labor Rights Forum, *http://www.laborrights.org/stop-child-labor/foulball-campaign/pakistan*
[34] Reebok Human Rights Report 2005. *http://www.reebok.com/Static/global/initiatives/rights/pdf/Reebok_HRReport2005.pdf*

identifies nine areas of its human rights efforts, two of which are child labor and forced labor.[35]

One of the primary ways that Monsanto addresses child labor is through first-phase producer training (suppliers of direct goods). In 2005, the company made sure that all of its partner contracts prohibit child labor. Monsanto distributed training materials to all hybrid cotton business partners and about 2,500 farmers. The training includes emphasis on Monsanto's policies that forbid the use of child labor. To start, Monsanto had to teach contractors about the negative effects of child labor, which many contractors did not understand, according to the company. In conjunction with the training, Monsanto created an incentive structure that helps the farmers to afford adult labor.

Monsanto cites a decrease in child labor: among the company's direct-goods suppliers, the percentage of children that make up the workforce dropped from 20 percent in 2004 to five percent in 2006.

However, some NGOs such as Stop Child Labour point out that the fight against child labor in the cotton seed industry still needs improvement. Stop Child Labour's *Out of work and Into School* report states, "prices paid to farmers are far too low, youngsters over the age of 14 (and adults) are made to work very long days and are exposed to pesticides, adults receive less than the minimum wage, labour unions are not involved in the Bayer/Monsanto initiative, and the schooling offered is at present (end 2007) insufficient. In addition full transparency about the implementation and impact of anti-child labour measures is lacking."[36]

Hewlett Packard

In 2001, Hewlett Packard issued its first Social and Environmental responsibility report. In 2003 it also formally adopted a human rights and labor policy. The 2003 report outlined HP's policy on child and forced labor as follows:

- **Freely-chosen employment.** Ensure no forced, bonded or involuntary prison labor is used in the production of HP products or services. Ensure that the overall terms of employment are voluntary.
- **No child labor.** Comply with local minimum age laws and requirements and do not employ child labor.
- **Freedom of association.** Respect the rights of workers to organize in labor unions in accordance with local laws and established practice.

Since 2003, Hewlett Packard has been actively involved with the United Nations Global Compact. HP used the Global Compact's nine principles to model its own human rights policies. The company states, "We will continue to review our policies and make adjustments as necessary to clearly communicate our support in these areas."[37] In addition, HP has taken the initiative to promote dialogue among other companies. In 2003, Hewlett Packard and Pfizer worked together to launch the Global Compact's North

---

[35] Monsanto website, "Issue Discussion: Child Labor in Agriculture", *http://www.monsanto.com/responsibility/our_pledge/stronger_society/child_labor.asp*
[36] Out of Work and Into School: Action Plan for Companies to Combat Child Labour, May 2008
[37] Hewlett Packard Global Citizenship Report for FY02, *http://www.hp.com/hpinfo/globalcitizenship/gcreport/downloads.html*

American Learning Forum, which brings together companies to discuss good practices for implementing the Global Compact's nine principles. Another HP project, the Central Europe Supplier Responsibility Project, targets multi-national companies that have small and medium-sized enterprises. The report, "Small Suppliers in Global Supply Chains" was the result of collaboration with the Danish Commerce and Companies Agency.

Tesco

In 2004, local activists in Uzbekistan started a campaign to urge countries to boycott Uzbek cotton that is harvested by children and exported all over the world. A few years later, several international companies responded. Among the companies that banned the cotton in their supply chain was Tesco, Wal-Mart, Target, Levi Strauss, Gap, Limited Brands and Marks and Spencer. Facing heightened and highly-critical international attention, the Uzbek government signed two ILO conventions against child labor.[38]

Steve Trent, Executive Director of the Environmental Justice Foundation, praised Tesco's decision to ban Uzbek cotton: "This ground-breaking move by Tesco – unprecedented from a major UK retailer – has the potential to change a multi-billion dollar industry."[39]

Tesco is based in the United Kingdom and is the world's third largest retailer. In a communication to its suppliers, Tesco wrote, "the use of organised and forced child labour is completely unacceptable and leads us to conclude that whilst these practices persist in Uzbekistan we cannot support the use of cotton from Uzbekistan in our textiles."[40]

Bon Appétit

Bon Appétit is a food service company that operates 400 university and corporate cafeterias in 29 states in the United States. The company prides itself on being sustainable and socially-responsible. However, in January 2009, Bon Appétit discovered that tomatoes it purchased were being picked by workers subjected to forced labor.[41] The Coalition of Immokalee Workers (CIW), a farm workers organization in Florida, alerted Bon Appétit to the fact that its tomatoes were grown in South Florida, an area with a history of migrant worker abuse. Since 1997, there were seven slavery convictions involving 1,000 workers in Florida.[42]

Bon Appétit executives then met with the CIW to discuss the company response. Fedele Bauccio, Bon Appétit's chief executive, insisted that "if no [grower] steps up, then I have to respond to my customers and not serve tomatoes."[43] Essentially, Bauccio proposed a boycott. However, CIW worried that a boycott would primarily just hurt the workers themselves. Although Bon Appétit buys almost 5 million pounds of tomatoes a

---

[38] ILRF, "We Live Subject to their Orders": A Three Province Survey of Forced Child Labor in Uzbekistan's 2008 Cotton Harvest, 2009.
[39] *http://www.ejfoundation.org/page483.html*
[40] *http://www.ejfoundation.org/pdf/Uzbekistan_Cotton%20Tesco_letter_to_%20suppliers.pdf*
[41] The Washington Post, "A Squeeze for Tomato Growers: Boycott vs. Higher Wages," Jane Black, April 29, 2009
[42] Bamco, "Game Changing Labor Standards"
[43] The Washington Post, "A Squeeze for Tomato Growers: Boycott vs. Higher Wages," Jane Black, April 29, 2009

year, other major companies buy many times more. Thus, CIW made the point that the company's boycott would not put a large grower out of business.

Bon Appétit and CIW reached an agreement that lays out acceptable working conditions and a code of conduct that forbids forced labor and physical violence or harassment. Bauccio agreed to increase wages, as CIW recommended, but also wanted to be sure that working conditions were improved. Though the agreement is considered a rough draft, both parties consider it a productive step toward furthering worker rights. In addition, Bon Appétit will reward growers that go beyond the minimum standards to improve working conditions even more.

4) Monitoring, Compliance and Certification

Gap (in partnership with Verité)

Gap's partnership with the non-governmental organization Verité was widely viewed as a success in improving Gap's social accountability. Verité provided Gap with monitoring services as well as helping to strengthen their Code of Conduct.[44] Verité, an independent, non-profit social auditing and research organization, has the mission to ensure "safe, fair and legal working conditions" for people around the world. To accomplish this, the organization promotes a multi-stakeholder collaboration and offers services that include monitoring, remediation, training, and research.

As a result of this collaboration, Gap won the Social Reporting Award for its *2003 Social Responsibility Report*. The 2003 report revealed the setbacks that Gap had experienced in trying to ensure compliance with its Vendor Code of Conduct. In particular, Gap cited areas in which, despite its efforts, monitoring did not actually capture all of the labor violations that probably had occurred. In one year, Gap revoked 136 factory approvals.

"Perhaps the most refreshingly honest aspect of the report is its admission that discrimination and freedom of association violations are likely more widespread than the data suggest. However, this is also one of the distressing aspects of the report, as it reveals how much work remains to lift the veil hiding these problems."[45]

Macy's (in partnership with Rugmark)

Macy's has partnered with Rugmark, an international non-profit that aims to end child labor in rug and carpet industry. Rugmark considers Macy's to be a "target company," one that it hopes will lead the way for other companies. With its Child-Labor-Free-Certification program, Rugmark uses certification to help consumers know if a rug was made using child labor and also funds educational programs for kids. The way it works is that each label has a serial number that traces the carpet back to the loom where it was produced. If the carpet manufacturer complies with Rugmark standards, they receive the Rugmark label.[46]

---

[44] Deborah Hirt, "Verite: Auditing Labor Standards", University of California San Diego. 2007.
[45] "Gap- Verité Collaboration Exemplifies Award-Winning Practice on Social Responsibility". SocialFunds.com. December 2004. *http://www.socialfunds.com/news/article.cgi/1581.html.*
[46] 2008 White Paper: Learning from the Rugmark Model to End Child Labor *http://www.rugmark.org/uploads/WhitePaper0409.pdf.*

With Macy's as a partner, Rugmark plans to document "a positive bottom line result for these trend-setters." The Macy's-Rugmark partnership is in line with Macy's effort since 1995 to enforce its Vendor/Supplier Code of Conduct. The Code of Conduct sets standards and requires all vendors to sign written affirmations, agreeing to comply with the Code of Conduct.[47]

Nike, Inc.

Data reporting in auditing and monitoring is becoming more common among apparel and footwear brands.[48] Nike is an example of a company that seeks to continually improve its monitoring and data reporting practices. In 2005, Nike was the first in its industry to voluntarily disclose the names and locations of the more than 700 contract factories that make Nike products.[49] The company believed the new disclosure and transparency would help promote compliance. According to the 2004 Report Review Committee for Nike's report,

> While noting that monitoring is not a sufficient or long-term solution to raising labor standards, the report presents Nike's extensive and evolving efforts to manage monitoring, integrate compliance into its business strategy through the Balanced Scorecard, and pursue multi-stakeholder initiatives that could lead to more systemic industry-wide improvements.

Nike contractors are required to have on file all documentation required for compliance with the Nike Code of Conduct, which covers child and forced labor. The contractors must make the documents accessible, and must agree to inspections whether or not they are given prior notice.[50]

In addition, like many other companies, Nike uses the Global Reporting Initiative (GRI) guidelines. GRI has produced a set of reporting standards called the *Sustainability Reporting Guidelines,* which are used on a voluntary basis and include environmental, social and financial issues. The section on child labor refers to the ILO's Minimum Age Convention (C. 138) and addresses issues of monitoring.[51] Still, some organizations like the Maquila Solidarity Network point out the limitations of data collection. In a 2007 report, the Network called for more improvement in the apparel industry:

> The collection, analysis and publication of hard data in company reports had the effect of concentrating management's attention on the problem and allowing for some measurement of progress. As that approach has become more widespread, the limitations of the methodology have also become clearer. The fact that updated data showed little or no progress in eliminating persistent abuses raised

---

[47] Macy's Web site, *http://www.macysinc.com/aboutus/sustainability/sweatshops.aspx.*
[48] Maquila Solidarity Network. The Next Generation of CSR Reporting: Will better reporting result in better working conditions? December 2007.
[49] The Corporate Social Responsibility Newswire, "Nike Issues FY04 Corporate Responsibility Report Highlighting Multi-Stakeholder Engagement and New Levels of Transparency", April 13, 2005.
[50] 2007 Nike Code of Conduct, *http://www.nikebiz.com/responsibility/documents/Nike_Code_of_Conduct.pdf.*
[51] ILO, Eliminating Child Labour: Guides for Employers: Guide Two: How Employers can Eliminate Child Labor, 2007.

serious questions about whether factory monitoring systems were effectively addressing the underlying causes of worker rights violations.[52]

Liz Claiborne

Liz Claiborne educates factory owners to emphasize that violations of labor standards is not economically beneficial in the long-run.[53] To ensure compliance, the company employs both internal and external monitoring. Liz Claiborne auditors carry out "spot inspections," for which contractors may or may not be given notice. For each supplier, internal auditors complete a human rights questionnaire, which requires detailed information on factory compliance with the Code of Conduct. The Liz Claiborne Workplace Code of Conduct is a set of standards that "prohibits child labor, forced labor and all forms of harassment or abuse, in addition to ensuring workers are paid the minimum wage, setting limits on overtime hours and requiring employers to honor workers' rights to freedom of association and collective bargaining."

The company's external monitoring is done by the Fair Labor Association. Two countries in particular where Liz Claiborne has independent monitoring are Guatemala and El Salvador. "We learned things from these two programs which has helped us improve our own internal monitoring programs… by participating in the FLA monitoring program, we are able to expand the independent monitoring concept to more countries, sometimes combining efforts with other participating companies."[54]

Target Corporation

Target clearly states that it "will not knowingly work with any company that does not comply with our ethical standards."[55] The company policy on child and forced labor is laid out in its Standards of Vendor Engagement: "No forced or compulsory labor. No child labor, which we define as being below the local minimum working age or age 14, whichever is greater. We make an exception for legitimate apprenticeship programs."

Since 1998, Target's Global Compliance team has worked to verify that Target's vendors comply with the company's Standards. These standards are laid out in the company's Vendor Conduct Guide, which include U.S. Customs and Border Protection laws and country-of-origin labor laws. One way that Target exerts control over its supply chain is by prohibiting vendors from subcontracting without Target approval. All subcontractors must be identified by Target:
"Vendors are required to register all factories used in the production of our merchandise (including subcontractors), maintain accurate and up-to-date information about each factory and authorize unannounced audits by Target team members or our accredited third-party auditors."

Furthermore, Target is a signatory of the National Retail Federation's "Statement of Principles on Supplier Legal Compliance," which it says is incorporated into its own standards and practices.[56]

---

[52] Maquila Solidarity Network. The Next Generation of CSR Reporting: Will better reporting result in better working conditions? December 2007
[53] Monitoring International Labor Standards: Summary of Domestic Forums, National Research Council, 2003.
[54] *http://www.lizclaiborneinc.com/rights/faqs.htm*
[55] Target Corporation Corporate Responsibility Report (2008)
[56] Target Corporation Corporate Responsibility Report (2006)

5) Remediation

Adidas-Salomon

When Adidas found underage workers in a factory in Vietnam, the company worked to address the problem and provide alternatives for the children in question. The problem in the factory was not just one or two cases: out of 2,000 workers, 200 were shown to be below the age required in the terms of employment (15 years old). Working with Vérité and an educational coordinator, Adidas developed a program that aimed to respond to the needs of the children and community. Not only was the factory required to pay the school fees of the children (under the age of 16), it had to continue to pay their working wages and guarantee a job once the children were out of school. Although it did not finance the program, Adidas did pay a quarterly advance for the factory's output, thereby relieving some financial constraints.[57] "Young" workers (ages 16 and 17) could continue work but with fewer hours and the average wage of the year prior. Education programs were set up on the factory premises.

Obeetee Ltd.

Obeetee is a leading carpet-exporting company that was founded in India in 1920. Today, it considers itself "a pioneer" in its approach to combating child labor. Obeetee's production occurs on more than 4,000 looms in about 1,000 Indian villages. The company states on its website: "Obeetee maintains the highest standards of workplace conditions and accountability. Our weavers and supporting manufacturers are amongst the best paid in the industry and receive superior benefits."[58]

During the 1980s, Obeetee implemented a "no child labor policy" in response to negative publicity and to new legislation from the Indian Government prohibiting the employment of children under 14 in carpet weaving.[59] Despite extensive monitoring, Obeetee is "aware that there are never any shortcuts, or instant solutions to solving such a massive problem such as existed with child labour in the Carpet Industry."

The company says it is committed to ensuring that children are not working for Obeetee or any other company. For this reason, Obeetee is involved with two key initiatives that provide services to children and their families. It helps to fund two organizations, the Carpet Export Promotion Council of India (CEPC)'s child welfare fund and to the Children Emancipation Society. Both organizations run child welfare programs, including schools that provide free education, monthly stipends, mid-day meals, vocational training, and health care to children.[60] A quarter percent from the sale value of each Obeetee carpet is donated to this CEPC Child Welfare Fund.[61]

CEPC states on its web site that the "Carpet Export Promotional Council of India is an autonomous body established to promote the export of Indian made carpets. CEPC

---

[57] ILO, Eliminating Child Labour: Guides for Employers: Guide Two: How Employers can Eliminate Child Labor, 2007.
[58] *www.obeetee.com*
[59] ILO, Eliminating Child Labour: Guides for Employers: Guide Two: How Employers can Eliminate Child Labor, 2007.
[60] "Child Labour Problem in the Unorganised Sector – The Obeetee Solution", International Finance Corporation, Labour Roundtable, October 9, 2001. *www.obeetee.com*
[61] Good Practice Note

is sponsored by the Ministry of Textiles, Government of India, but its services are funded primarily by its members. Compliance with the CEPC's Child Labour Code of Conduct, is Mandatory for all Carpet Exporters."[62]

Cadbury

Cadbury began the Cadbury Cocoa Partnership in January 2008 with the goal to have "thriving rural communities that support a sustainable cocoa supply chain".[63] In just one year, the program spread to 100 communities in Ghana. Cadbury reports: "The 100 communities who have now joined the partnership have been identifying their main development needs, including the construction of new school buildings or forming Cocoa Youth Clubs to encourage the next generation to remain with agriculture, particularly cocoa farming."

Cadbury's focus is education and training. The company partnered with Digital Links International to train teachers in technology. Farmers themselves also receive education through programs that help to improve the quality and quantity of farmer yields. In addition, Cadbury encourages young people to start businesses or learn other skills. Young people can receive enterprise loans or apprentice with a tradesperson with the financial support provided by Cadbury.

Bonita

In 2002, international pressure prompted the banana industry to try to improve working conditions on Ecuadorian plantations. Human rights advocates criticized the suppression of workers unions on the plantations. Advocates also called for the eradication of child labor. That year, it was estimated that at least 6,000 children were working on large plantations in Ecuador.[64] That number did not include the many small plantations across the country.

The Banana Social Forum was implemented in 2003. Bonita, the fourth largest global banana exporter, is highly involved in the Forum. Sponsored by the Ecuadorian government, the Forum has the goal to develop and implement initiatives that aim to eradicate child labor in the banana sector. For children between the ages of 15 and 18 legally working on the farms, the Forum aims to improve their living conditions and educational opportunities. The Forum's representatives are from the Ministries of Labor and Agriculture, CORPEI, INNFA, producers, exporters and workers in the banana sector, and UNICEF. The ILO became involved with the Forum to develop a "tripartite initiative based on social dialogue with effective trade union participation."[65]

---

[62] Carpet Export Promotional Council of India, *http://indiancarpets.com/welfare.htm*.

[63] *http://www.cadbury.com/ourresponsibilities/cadburycocoapartnership/Pages/cadburycocoapartnership.aspx*.

[64] The New York Times, "In Ecuador's Banana Fields, Child Labor is Key to Profits" *http://www.nytimes.com/2002/07/13/world/in-ecuador-s-banana-fields-child-labor-is-key-to-profits.html*.

[65] International Business Forum on Engaging Business- Addressing Child Labor Case Studies, February 25, 2009, Atlanta, GA.

APPENDIX F

Since 2003, Bonita has established 8 schools, 345 houses with basic services for workers, 6 health care units, and 6 commissaries. The Forum works to raise awareness on child labor amongst the trade unions, entrepreneurs, and the families and children themselves. Other Forum activities include a child labor inspection and monitoring system and encouraging banana companies to agree to labor inspections on their farms and plantations.

Appendix G
Submissions to the Workshop

Prior to the meeting, experts from a variety of organizations were invited to submit comments on the framework. Several submissions were received and they are reproduced here.

1. As You Sow
2. Cadbury
3. Monique Oxender, Global Manager, Supply Chain Sustainability, AIAG Sustainability Loan, Ford Motor Company
4. Chisara Ehiemere, Transfair, USA
5. CREA, Inc.

APPENDIX G

1. As You Sow
Actions Companies can take to Address Forced and Child Labor in Uzbekistan

The following recommendations outline available options for companies to implement to assist in transforming Uzbekistan's cotton sector.

1. Acknowledgement

Regardless of whether or not a company has identified Uzbek cotton in its supply chain, a viable first step is making a public statement renouncing the actions of the Uzbek government with regard to forced and child labor in the cotton sector. This could take the form of a press release, media report and/or announcement on the company's website.

2. Identification

Making the decision to not use Uzbek cotton is contingent upon identifying its presence in the first place. There are several ways of doing so:

   a. Communicate with suppliers that your company is concerned with sourcing Uzbek cotton and is currently reviewing its policies on the issue (see Sample Company Language)
   b. Ask suppliers to include country of origin information for the materials that go into their sourced products on all Textile Information Sheets (see Marks & Spencer example)
   c. Review previous Bills of Lading and product specification sheets to retroactively identify Country of Origin
   d. Employ tracing mechanisms such as Historic Futures *String* program
   e. Begin mapping the social and environmental footprint of your cotton procurement practices
   f. Implement an internal tracking system for all products (preferably online and via barcodes)

Those companies that find they are not sourcing Uzbek cotton should publicly denounce the practice, and introduce a ban until the government has made efforts to correct this ongoing injustice. Those companies that identify Uzbek cotton in their supply chains should take immediate steps to procure cotton from other countries.

3. Engagement and Education

Companies should engage shareholders, MSIs (Multi-stakeholder Initiatives), NGOs, trade associations and other stakeholders to coordinate efforts and strategic plans. Further, correspondence with the Government of Uzbekistan, own domestic government, and international organizations such as the ILO, UNICEF and the World Bank will raise the profile of the issue and help exert more pressure upon the Uzbek government.

Further, companies that have taken the aforementioned steps could encourage additional companies to address this issue as they deem appropriate. Through industry associations, MSI gatherings or trade shows, this issue is an industry-wide concern that must be addressed in an inclusive manner. Sharing best practices and experiences on how your company was able to take these steps will make it less cumbersome for those wishing to pursue a similar strategy.

4. Summary

Below are three basic categories that should give an indication of where a company fits in comparison to other companies taking similar steps toward supply chain transparency and traceability generally, and in addressing sourcing Uzbek cotton specifically.

Basic

- Acknowledge problem of forced labor in the cotton industry globally
- Acknowledge problem of systemic forced labor specifically in Uzbekistan's cotton industry
- Publicly state that company is going to look into whether or not it is sourcing Uzbek cotton, and will take further action depending on conclusion of investigation
- Initiate cotton footprint mapping (social and environmental)
- Engage with shareholders, industry associations, MSIs, and human rights activists

Intermediate

- Communicate with own domestic government, government of Uzbekistan, and international institutions (USCIB/ILO, World Bank, ICAC, UNICEF) about concerns
- Trace chain of custody (internally or through suppliers)
- Formal Supplier communications

Advanced

- Country of origin of cotton/fiber required on Textile Specification Sheets
- Internal system for tracking and reporting
- Trace cotton source back to originating farm
- Exert influence inside the ILO, with the government of Uzbekistan

Update on Forced & Child Labor in Uzbekistan

As in past years, last fall state officials ordered up to 2 million children in Uzbekistan, aged 11 to 17, to leave school to work 10 hours a day, 7 days a week, under hazardous conditions harvesting cotton for two months to fill government-mandated quotas. A diverse group of stakeholders have come together to address the situation.

Update from the Field

- Approximately two thirds of schools are subject to compulsory recruitment of children between ages 11 and 17.[1]

- Children are in the cotton fields for a total of 51 - 63 days without weekend breaks and under detrimental sanitary, health and nutritional conditions.[2]

- School administrators used methods of physical abuse and public shaming to force children to meet daily quotas, which was 30-60 kg of cotton depending on age. In one case such treatment resulted in a suicide.[3]

- Parents who resisted sending their children to pick cotton were threatened by the authorities and risked having their welfare benefits and utility services withheld.[4]

- Each year scores of children are injured or killed in the harvest due to the lack of safety measures and adult supervision. In the fall of '08 alone, there were five reported fatalities.[5]

Activities by Companies

- Eight companies have written to their suppliers, are taking measures to exclude Uzbek cotton from their supply chain, and/or are starting to trace the country of origin of the cotton they are using.

- 15 brands and retailers are partnering with other stakeholders on this issue by participating on conference calls, attending meetings and issuing public statements.

- In 2008, articles referencing company actions appeared in the Financial Times, Fortune Magazine, Reuters, Dow Jones, and Just-Style.

Actions by Stakeholders

- The U.S. Generalized System of Preferences (GSP) complaint filed by the International Labor Rights Forum in 2007 was held open for continued review through 2008.

- The U.S. State Department convened a meeting in May 2008, which was attended by 48 people from Trade Associations, Brands and Retailers, Social and Faith-Based Investment Firms, International Institutions, Human Rights Groups, and U.S. Government Agencies.

- Joint Trade Associations letters and joint Investor-NGO letters were issued to President Karimov, Former U.S. Secretary of State Rice, and International Labor Organization (ILO) Director General Somavia in August 2008.

- A panel of experts created by the International Cotton Advisory Council (ICAC) issued a Literature Review and Research Evaluation relating to *Social Impacts of Global Cotton Production* where the issue of forced child labor in Uzbekistan was acknowledged.[6]

- The International Organization of Employers (IOE) submitted an ILO complaint regarding forced labor in Uzbekistan in November 2008.

- Starting in September 2008, a group of U.S. investors started contacting cotton merchants and brokers, and engaging the CEO of the Dubai Multi-Commodity Center (DMCC).

- During the Universal Periodic Review in December 2008 (a new mechanism introduced by the UN Human Rights Council), nine states raised their concerns about the practice of forced child labor in Uzbekistan.

- A small U.S. multi-stakeholder group met with the Ambassador of Uzbekistan in December 2008 requesting the ILO be invited into Uzbekistan; a response is pending.

---

1. "Still in the Fields," Environmental Justice Foundation, October 2008. http://www.ejfoundation.org/page341.html
2. Ibid.
3. Child Labor in Fall 2008 Uzbek Cotton Harvest, International Labor Rights Forum, November 2008 http://www.laborrights.org/stop-child-labor/cotton-campaign/resources/1843
4. Ibid.
5. Ibid.
6. ICAC, July 2008, http://www.icac.org/seep/documents/english.html

Government of Uzbekistan Compliance with International Instruments

The Government of Uzbekistan (GOU) ratified two ILO conventions, the Worst Forms of Child Labor (Convention 182) and Minimum Age (Convention 138); however, it has not yet complied with the standards laid out by these conventions. C182 was ratified in June of 2008, but before going into effect one year later as per the usual process, multiple steps such as a comprehensive survey and a listing of affected sectors are necessary to be rendered "active".[1] As of March of 2009, no effort to complete these requirements has been initiated by the GOU. For a 2008 request for information from the ILO Committee of Standards on Forced Labor, the GOU has yet to provide the needed information. C138, the Convention on Minimum Age, was finally deposited appropriately in March 2009— eight months after its original submission— so it is now recognized as "ratified" by the ILO.

The GOU Cabinet of Ministers passed a resolution to adopt a National Action Plan (NAP) on September 12, 2008 to monitor the implementation of ILO Conventions 138 and 182.[2] Thus far, no documents on how the NAP will be implemented have been issued. The GOU has repeatedly asserted that it has banned forced child labor. However, reports suggest it continued to require children and other categories of youth and adult populations, including elders and housewives, to pick cotton during the fall '08 harvest.[3]

---

1. "ILO Core Labor Standards," International Labor Organization. 2008. http://www.ilo.org/ilolex/english/docs/declworld.htm
2. "Invisible to the World." The School of Oriental and African Studies, University of London. 2008. http://www.soas.ac.uk/cccac/centres-publications/
3. "ILO Uzbekistan update, November 2008." International Labor Rights Forum. November 2008. http://www.business-humanrights.org/Links/Repository/670344/link_page_view

Further Information

Additional opportunities to learn more about the issue and share best practices by companies will be available throughout 2009 via seminars, webinars, and conference calls. To be informed of future activities or for more information, send an email to: cotton@asyousow.org (this is not a list serve).

Companies Taking Action on this Issue

American Eagle Outfitters	Marks & Spencer
C&A	Nike, Inc.
Columbia Sportswear	Patagonia
Gap Inc.	Target Corporation
Hanesbrands, Inc.	Tesco
Jones Apparel Group	The Walt Disney Company
Levi Strauss & Co.	Wal-Mart Stores, Inc.
Limited Brands Inc.	

Collaborating Industry Associations

- American Apparel and Footwear Association (AAFA)
- International Organization of Employers (IOE)
- National Retail Federation (NRF)
- Retail Industry Leaders' Association (RILA)
- US Association of Importers of Textiles and Apparel (USA-ITA)
- US Council for International Business (USCIB)

Additional Resources

- BBC News Night Story - http://news.bbc.co.uk/1/hi/programmes/newsnight/7068096.stm
- The International Labor Rights Forum (ILRF) Cotton Campaign - http://www.laborrights.org/stop-child-labor/cotton-campaign
- Environmental Justice Foundation (EJF) Cotton Campaign - http://www.ejfoundation.org/page141.html
- Photos - http://www.iwpr.net/galleries/centasia/grabka/01.html

3/12/09

Compiled by the As You Sow Foundation, cotton@asyousow.org

Sample Language Regarding Uzbek Cotton

Company Statement on Website

Code of Conduct for Uzbekistan

For all C&A suppliers the C&A Code of Conduct is binding. It strictly states that child labour is unacceptable for C&A.

Already in December 2007 C&A obliged all of its global suppliers in written form not to use cotton fibre from Uzbekistan in the manufacturing of C&A products. Furthermore, C&A has requested all suppliers worldwide to clearly state the origin for cotton fibre which is used within any C&A merchandise.

Source: http://www.c-and-a.com/aboutUs/socialResponsibility/

Company Letter to Suppliers

Dear Supplier,

As you will be aware worker welfare is of paramount importance to Tesco Stores Limited, which is why we go to great lengths to conduct ethical inspections in all factories supplying Tesco, wherever they are in the world.

Following ongoing discussions with campaign groups on the subject of cotton production and the use of child labour in this part of the supply chain, we feel the need to re-iterate Tesco's revulsion at the use of child labour. We realise that child labour is a complex issue with many causes which we acknowledge are hard to effect individually. However the use of organised and forced child labour is completely unacceptable and leads us to conclude that whilst these practices persist in Uzbekistan that we cannot support the use of cotton from Uzbekistan in our textiles.

We understand that cotton is an internationally traded commodity and that raw cotton sources are not always easily identifiable. However from AW08 onwards we will require you, wherever possible, to identify the source of raw cotton used in Tesco textiles products and document this. We will reserve the right to randomly audit records to monitor the source of raw cotton. Where it has not been possible for you identify the source of raw cotton we will require you to advise the relevant Technical Manager of this and the reasons that this could not be done ahead of starting production.

For further information regarding conditions in the Uzbekistan cotton industry please refer to: http://www.ejfoundation.org/pdf/white_gold_the_true_cost_of_cotton.pdf and for advice on tracking the sources of raw cotton please contact Abi Rushton.

Please return a signed copy of this letter as an indication of your commitment to this initiative to Nikki Gilman by December 31st 2007.

Yours sincerely,

Terry Green
CEO Tesco Clothing and Hardlines

Source: http://www.ejfoundation.org/pdf/Uzbekistan_Cotton%20Tesco_letter_to_%20suppliers.pdf

Company Letter to Vendors and Mills

August 26, 2008

To our Gap Inc. Vendors and Fabric Mills:

We are writing to you regarding the use of cotton from Uzbekistan. As you know, the issue of child labor is one that is very concerning to Gap Inc. And continues to be on the minds of our customers. We have all seen reports alleging that children are working in cotton fields in Uzbekistan under hazardous conditions in violation of international labor standards and children's basic human rights.

Gap Inc. is against the use of child labor in any stage of the production of our apparel.

Our Code of Vendor Conduct includes strict prohibition the use of child labor.

Violations of our code can result in the termination of our contract and business relationship.

It is our expectation that vendors with which we contract not knowingly source textiles from mills sourcing cotton from Uzbekistan. We have recently communicated this expectation to our vendors, and wanted to be sure we shared this position with the fabric mills our vendors source from as well.

Thank you in advance for your ongoing attention to this matter. We appreciate your support.

Sincerely,

Dan Henkle
Senior Vice President
Social Responsibility

Stanley P. Raggio
Senior Vice President
Gap International Sourcing

Source: http://www.gapinc.com/public/SocialResponsibility/sr_enviro_design.shtml

Company Letter to Suppliers

Marks and Spencer: Cotton Fibre Sourcing from Uzbekistan

There have been ongoing concerns regarding the use of government backed forced child labour during the cotton picking season in Uzbekistan.

We have been working with our textile suppliers over the past few months to firstly understand where Uzbekistan cotton (world 3rd largest cotton exporter) may be used in M&S garments.

Having done this review we are now specifying that our suppliers MUST NOT use any cotton fibre or fabric sourced from Uzbekistan. This will remain our position until such time that there is clear evidence of a change in the Uzbek cotton industry.

Cotton supply chains from field to retail are complex and lengthy. It has often been considered difficult to operate systems that confirm country of origin for all cotton fibre used in M&S products. However in light of the Uzbekistan situation we need to incorporate fibre country of origin (cotton only) into our buying specifications. This will be done through modification of the fabric technical submission document with immediate effect.

We will, of course, work with suppliers in resolving any issues that may arise re Uzbekistan but equally we need to act now.

Our position regarding Uzbekistan is part of an overall cotton sourcing strategy which is close to completion. We will advise further on our plans.

Yours sincerely,

Graham P. Burden
Sustainable Textiles and Cotton Specialist

Company Public Statement

TARGET Media Statement

As a retailer, we have an obligation to ensure that the products we carry in our stores are made legally and ethically. We maintain Standards of Vendor Engagement and conduct regular, random unannounced audits to ensure compliance with our protocols.

In keeping with our commitment to ensure safe and healthy working conditions and respect for the rights of the people who work in factories that manufacture our products, Target does not knowingly buy or sell products that use cotton sourced from any country that condones the use of forced child labor. To the best of its knowledge, Target currently does not source any apparel or home products from Uzbekistan nor does it use Uzbek cotton in its textiles used to manufacture those products from other countries.

We are informing our business partners of our concerns and are requesting they do not use cotton sourced from any country with a known record for forced child labor. We are sharing this position appropriately with the U.S. Government, Non-Governmental Organizations and other industry leaders who have approached us with a concern regarding the use of forced child labor in cotton production in Uzbekistan.

Company Press Release

Wal-Mart Takes Action to End Forced Child Labor in Uzbekistan

BENTONVILLE, Ark., Sept. 30, 2008 – Wal-Mart Stores, Inc. has instructed its global supply base to cease sourcing cotton and cotton materials from Uzbekistan in an effort to persuade the Uzbek government to end the use of forced child labor in cotton harvesting. This action follows months of work with industry trade associations, government agencies, non-governmental organizations and socially responsible investment groups to form a common position in condemning the Uzbek government's practices.

"We have formed an unprecedented coalition, representing 90 percent of the U.S. purchases of cotton and cotton-based merchandise, to bring these appalling child labor conditions to an end," said Rajan Kamalanathan, vice president of ethical standards. "There is no tolerance for forced child labor in the Wal-Mart supply chain."

With Wal-Mart's active participation, four industry trade groups, the American Association of Footwear and Apparel, Retail Industry Leaders Association, National Retail Federation, and the United States Association of Importers of Textiles and Apparel sent a joint letter to the Embassy of Uzbekistan on Aug. 18, 2008, demanding an immediate end to the use of forced child labor in cotton harvesting. In response, the Uzbek government issued on Sept. 12, 2008 a National Action Plan which details steps to eradicate the use of child labor. Once these steps can be independently verified, Wal-Mart will modify the direction to its suppliers.

Source: http://walmartstores.com/FactsNews/NewsRoom/8637.aspx

Company Response to Investor Letter

LEVI STRAUSS & CO.

LEVI'S®
DOCKERS®
LEVI STRAUSS SIGNATURE™

LEVI STRAUSS & CO.
1155 BATTERY STREET
SAN FRANCISCO, CA 94111

Dear Patricia:
April 23, 2008

Thank you for your letter outlining the concerns of As You Sow and others in the socially responsible investment community about human rights abuses in the cotton fields of Uzbekistan. We welcome your call for a collaborative, nuanced approach to address forced child labor in cotton harvesting in Uzbekistan.

Levi Strauss & Co. (LS & CO.) is firmly committed to sourcing in countries respectful of human and worker rights. In fact, this commitment is embedded in our comprehensive sourcing guidelines that were established to help us source in countries and with business partners that follow workplace standards and business practices consistent with our company's values.

Consistent with our commitment to ensure that the people making our products are treated with dignity and respect and work in safe and healthy conditions, LS & CO. will no knowingly use cotton sourced from Uzbekistan in the production of our products until there is clear evidence that action is being taken to eliminate the use of forced child labor in the Uzbek cotton industry. Since 2003, LS & CO. has prohibited the use of textiles from Uzbekistan. We have also prohibited the sourcing of LS & CO. apparel from Uzbekistan.

To achieve this commitment, we will inform our suppliers of our concerns and request that they not use cotton sourced from Uzbekistan or textiles produced using Uzbek cotton. We will share this position and associated actions with the U.S. Government, nongovernmental organizations, and other industry leaders who have approached us with an interest in forced child labor in cotton production in Uzbekistan. We will also participate in the May 21 U.S. Department of State multistakeholder forum and look forward to any guidance or opportunities for future engagement that it might yield.

If you have any questions about our position, please contact me. We look forward to future discussions and collective multistakeholder action to address this issue.

Sincerely,
Michael Kobori
Vice President, Supply Chain, Social and Environmental Sustainability

Compiled by the As You Sow Foundation. cotton@asyousow.org

APPENDIX G

2. Cadbury

Version 5
24/06/08

CADBURY COCOA PARTNERSHIP VISION INTO ACTION - GLOBAL

Vision	Thriving rural communities that support a sustainable cocoa supply chain
Governing Objective	Cocoa-growing communities empowered to take leadership in: Meeting their long-term goals and delivering sustainable cocoa production
Delivery Approach	Our approach puts the community first, works through partnerships and builds local capacity, promoting community-centred activities delivered at scale through policy advocacy and reform, innovation and research

Indicator Scorecard	Communities empowered	Productivity in target communities	Household income in target communities	Capacity of key local and national institutions	Livelihood opportunities for rural youth	Biodiversity and Reduced Deforestation Rates	Key Health and Basic Education Outcomes

Strategic Themes	SUSTAINABLE LIVELIHOODS FROM COCOA	SUSTAINABLE LIVELIHOODS FROM OTHER MEANS	COMMUNITY CENTRED DEVELOPMENT	INSTITUTIONAL ENGAGEMENT

Key Activities	1. Farmer organisation
	2. Advocacy with other stakeholders to support cocoa farming
	3. Farmer training
	4. Access to extension services
	5. Improved farming and processing techniques
	6. Improved productivity and farm efficiency
	7. Improved farmer incomes
		1. Alternative income sources (local and external markets)
		  - research into options
		  - agro-processing
		2. Skills training including financial and entrepreneurial skills
		3. Access to credit and rural banking
		4. Youth engagement
			1. Community mobilisation leading to community mg'mt and planning
			2. Maintain/enhance natural environment
			Access to:
			3. Education
			4. Healthcare
			5. Energy sources
			6. Potable water
			7. Community services and technology
				1. Institutional synergy (traditional, district, national and int'l)
				2. Identify/work to strengthen weak institutions
				3. Organization and individual capacity building and clarification of roles and responsibilities of respective institutions

Cross cutting themes:
Addressing HIV
Addressing Gender, Discrimination and Diversity Issues
Addressing the worst forms of child labour and trafficking
Biodiversity Conservation and Environmental sustainability

3. Monique Oxender, Global Manager, Supply Chain Sustainability, AIAG Sustainability Loan, Ford Motor Company

• Questions around impact and relevance need to take into consideration whether the project began with a root cause analysis including the entire system contributing to the abusive situation. Consequently, did action result in a change to the system contributing to the abuses rather than a band-aid? Reduction of these labor abuses is not easy to measure and should not be a single point in time measurement. Also included should be a rate of recurrence over x amount of time to account for cyclical behaviors. Take the example of forced labor in Brazil. One out of every 10 temporary work opportunities may result in a forced labor situation but due to the need for employment and income, lack of education as well as lack of interest in more permanent employment, laborers are willing to take that chance. And this cycle plays out over a period of months to years.
• I am unclear as to how program/practice effectiveness differs from impact.
• Sustainability questions should specifically require consideration of funding and stakeholder involvement. Actually, no where do I see stakeholders mentioned and I think this is an essential aspect to understand with projects such as this - both who is involved and how.

Hope this helps. Looking forward to the discussion.

4. Chisara Ehiemere, Transfair, USA

Fair Trade Certification includes compliance criteria related to child and forced labor for both small farmers and hired labor situations, an addition to numerous other criteria for social, socioeconomic, and environmental development. As such, any impact, or goals of the program would not be limited to child labor or forced labor elimination, but to a much broader set of goals. That said, this is an important part of the standards, and discovery could lead to suspension or decertification.

The standards are written to both recognize that small farmers may have their children perform some tasks on the farm, but these must be limited, and cannot interfere with schooling. For hired labor situations, discovery of child or forced labor cannot simply be "fixed" by ceasing employment, but also trying to ensure that these workers are not forced into worst forms of labor situations.

In terms of replicability, the compliance criteria that are used are the same worldwide. Replicability can best be achieved by figuring out what types of labor situations a program wants to work with (small farms, cooperatives, large farms, small factories, large factories etc), and different levels of pervasiveness of these types of labor situations, and working out a compliance criteria for each type. For example, Auditor training materials and audit methodology may vary slightly by country/region based on local practices, and the type of labor set-up being reviewed.

For Fair Trade certification, the practices must continue in order to retain the certificate. I would imagine that now that sensitivity and awareness to the issues exists, there may be some that would continue to comply, but in most cases, audit and certification is necessary to ensure that responsible choices continue to be made.

Cost effectiveness is difficult to separate out because the audit is a comprehensive review of all compliance criteria, not just criteria related to child/forced labor. I imagine, however, that one might look at verification methods that include, for cooperative structures, internal control systems and scientific sampling techniques. For 100% certainty, you would almost have to have an auditor on the ground 100% of the time, but there are sampling techniques that can give a high level of certainty.

5. CREA, Inc.

CREA Inc.
P.O. Box 2507
Hartford, CT 06146-2507
TEL: 860.527.0455
FAX: 860.216.1072
e-mail: crea-inc@crea-inc.org
Web site: www.crea-inc.org

Center for Reflection, Education and Action, Inc.

May 1, 2009

Draft Criteria – with suggested revisions

I. Establishing baseline information
1. How was the baseline information collected?
2. How did you measure reliability of the baseline data?
3. What was the geographic area for the baseline data?
   a. What is regional?
   b. National?
   c. Local?
D. How recent was the data collection?
E. Did the data collection distinguish between child labor and forced labor?
F. Did the data collection distinguish by gender?

II. What are the specific program goals?
   A. How were the goals established
   B. Are they measurable
      a. Quantitatively
      b. Qualitatively
   C. How often will progress on program goals be measured?
   D. By whom will the progress be measured?

III. What are the specific program components?
   A. How were the components designed?
   B. How distinguishable were the separate components in terms of measurability?
   C. How are/were the effects of each component measured?

IV. Relevance – at start of project
   A. Is there a set of assumptions about how activities will lead to outcomes?
   B. Do you understand why the program works?

APPENDIX G

    C. Is there a logical connection between the inputs, activities and expected outcomes?

V. Consequences
    A. How are you distinguishing between impacts and consequences?
    B. What were the direct consequences of the program?
    C. What were the indirect consequences of the program?
    D. How did you measure the consequences?
        1. Quantitatively?
        2. Qualitatively?
    E. What were the unintended consequences of the program?
    F. What were the direct impacts of the program?
    G. What were the indirect impacts of the program?
    H. How did you measure the impacts?
        1. Quantitatively
        2. Qualitatively

VI. Impact(s)
    A. Did the overall program reduce child labor?
    B. Did the overall program reduce forced labor?
    C. Did the program benefit child laborers?
    D. Did the program benefit forced laborers?
    E. What were these specific benefits?
    F. How were they documented?
    G. How were they measured?

VII. Did the program achieve its goals?
    A. How was this determined
        1. Qualitatively
        2. Qualitatively
    B. Could the program sustain its goals over time?

VIII. Relevance – at completion of project
    A. Were the starting assumptions about how activities would lead to outcomes met?
    B. Do you understand why the program worked?
    C. Is there a logical connection between the inputs, activities and expected outcomes? Can these be documented?

IX. Sustainability
    A. Is the practice likely to continue (as needed)? Why or why not?
    B. Is the benefit likely to continue effectively?
    C Does the institutional capacity necessary to sustain the benefits and/or exist?
    D. Does the will to sustain these practices exist?

E. What is necessary for local ownership of the program or practices to continue?

X. Replicability
   A. Could the practice be implemented with modest adaptation in other settings?
   B. What factors limit replicability?
   C. What factors encourage replicability?

XI. Cost effectiveness
   A. How is cost effectiveness measured?
   B. Were the benefits sufficient to warrant the cost in terms of money and time?
   C. Were the benefits adequate in relation to likely benefits from comparable investments?
   D. How was cost effectiveness measured?
   E. Can the business case be made for the project or work as a means of encouraging others to do the same work?

Appendix H
Submissions following the Workshop

At the end of the workshop, presenters and participants were invited to submit further comments. One further submission was received from Donna Chung who indicated that the "views presented are my own and do not necessarily reflect the views of the ad hoc [planning] committee as a whole, or those of Sandler, Travis, and Rosenberg, P.A. Dr. Chung's comments are included verbatim in this appendix.

Criteria for Identifying Promising Business Practices to Eliminate Child & Forced Labor: Three Foundational Elements

Donna E. Chung, Ph.D.

Submitted to U.S. Department of Labor
Office of Child Labor, Forced Labor & Human Trafficking

July 7, 2009

During the May 11-12, 2009 National Academy of Sciences (NAS) Workshop, a diverse group of stakeholders provided comments on the framework under development for identifying and organizing business practices to eliminate forced or child labor in the production of goods. The "Draft Criteria," intended to contribute toward the development of the framework, reflected a small portion of the Ad Hoc Committee's[1] earlier reflections on the topic. It was circulated prior to the Workshop and provided a starting point for the Workshop's discussions. This document expands on the "Draft Criteria," taking into account the Workshop discussions, as well as my own deliberations on the subject.

For the purposes of this document, an important distinction is made between the terms "framework" and "criteria".[2] Consistent with the Department of Labor's "Statement of Task" for the NAS, "framework" refers to the conceptual structure to be used for both (1) *identifying* and (2) *organizing* pertinent business practices. "Criteria," in contrast, refers more narrowly to the evaluative tool that would be used, primarily, to achieve the first of the objectives of *identifying* eligible practices. Criteria are a set of standards (or indicators) that could be applied to distinguish "eligible" functional outputs from "ineligible" ones – in this case, "promising" business practices for reducing child/forced labor from not-so-promising practices. This document focuses on the first task, then, of commenting on the criteria to be used for identifying promising practices.

Many good insights were shared at the two-day Workshop pertaining to the elements to be included in the criteria. For example, there was a general consensus reached around the need for the criteria to address such important issues as:

- Effective problem analysis, including root-cause analysis;

---

[1] Ad Hoc Committee on Approaches to Reducing the Use of Forced or Child Labor
[2] I believe the use of these terms interchangeably at the Workshop created some level of confusion for the participants and deserves clarification.

- Community stakeholder engagement;
- Contextualization of the practice to local milieu;
- Institutional capacity building;
- Robust evaluations & impact assessment; and
- Sustainability.

What was missing in the discussion, however, was a consensus on how these disparate and important issues might be grouped into categories that provide a structure for analysis and examination. I submit the following three categories as a structural foundation for the criteria. They are not meant as an exhaustive list of items to be included in the final criteria. Rather, they constitute a suggested means of organizing the long list of valuable ideas and concepts shared at the Workshop. I hope that they contribute toward building a solid conceptual foundation for the criteria that will be used to identify good practices in this area.

(1) The first category of criteria has to do with the relationship between the particular practice and the business entity responsible for the practice. This category has to do with the question: For a practice to be considered "good", what qualities should it exhibit with regard to its relationship with the business entity? What place should the particular practice have within the whole of the operations and ethos of the business entity in question? Here, my argument is that a set of criteria should point to practices that are holistic and integrated – practices that demonstrate that the objective of eliminating child/ forced labor is integrated into the decision-making processes of every stage of the business' value chain, from product design and engineering, to raw material selection, manufacturing, to sales, marketing, and product decomposition.

(2) The second category has to do with the relationship between the particular practice and the target beneficiaries. The question addressed here is: For a practice to be considered "good", what qualities should it exhibit in the way it relates to the target beneficiaries? Here, the criteria are meant to point to practices that demonstrate in their problem analyses, intervention-design, implementation, and evaluation an emphasis on target community participation and the extent to which the root causes endemic to the community are being addressed by the practice intervention.

(3) The third category has to do with a set of criteria that point to a robust system of internal and external evaluations and effective management of data and information. The question here is: For a practice to be considered "good", what qualities should it exhibit in the way it utilizes external and internal evaluation methods?

The diagram below depicts the relationship between the three categories.

APPENDIX H

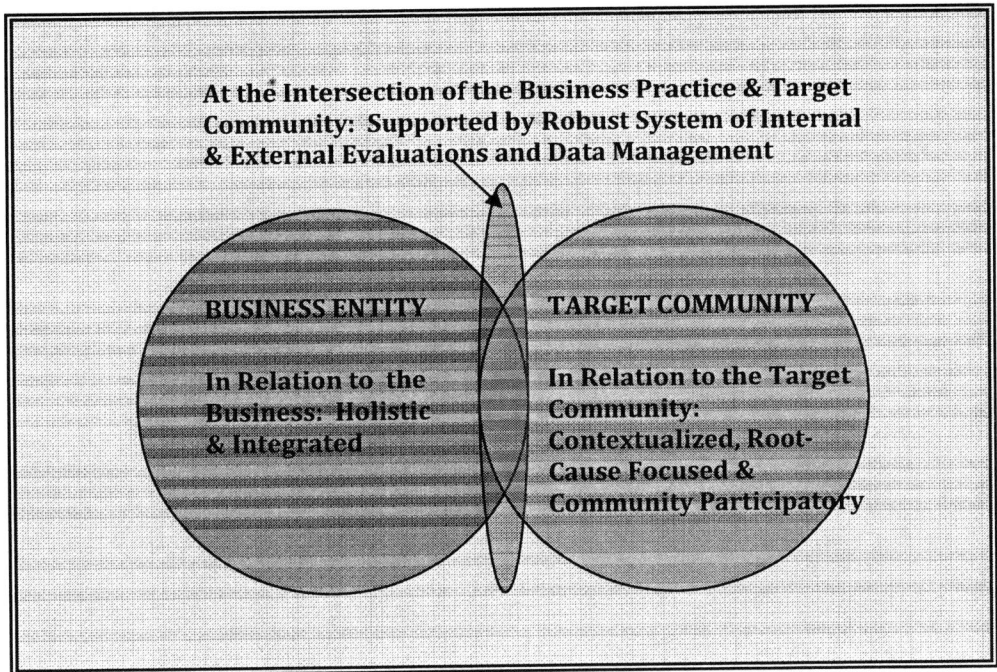

The Blue circle (on the left) represents the Business Entity, for whom the particular practice/intervention should be integrated into its entire business process. The Red circle (on the right) represents the Target Community, whose participation should shape and influence the design, implementation, and evaluation of the particular practice. The Green circle, which encompasses the overlapping portion of the other two circles, represents the evaluative work that should be carried out to measure the effectiveness of the intersection between the business' practice and the community. The following sections elaborate on these categories of criteria for identifying good practices.

I.   Category 1: Integrated & Holistic in Relation to the Business

Overarching Criterion: With regard to the practice's relationship to the business, the practice (or intervention) should demonstrate that the objectives of eliminating child/forced labor are integrated into the whole of the business' operational principles and practices. This includes the integration of the objective of eliminating child and/or forced labor into the decision-making processes of every stage of the business' value chain, from product design and engineering, to raw material selection, in-bound logistics (e.g., procurement, sourcing, pricing, etc.), core operations (e.g., manufacturing and production), out-bound logistics (e.g., packaging, distributing, transporting, etc.), marketing and sales (e.g., pricing practices, consumer information, etc.), and product teardown (e.g., recycling, product decomposition).

Significance of Such Holistic & Integrated Approach

Problems arise when business practice to eliminate the use of exploitative labor is *not* integrated into the whole of its business operations. When an intervention or practice (such as the development of codes, guidelines, and monitoring systems) is developed and implemented in isolation from the rest of the company's operations, it is likely that it lacks:

1. *Genuine Commitment*: When a business cannot demonstrate that the practice is integrated into the whole of its operations, this is often a sign that there is not a genuine long-term commitment to the objectives of the practice.

2. *Effectiveness*: When the objectives of a practice are not integrated into all stages of business operations, conflicting internal practices may arise and reduce the likely effectiveness of the "good" practice. For example, even if sound codes and manuals are developed and distributed to suppliers, if such practice is not matched by appropriate purchasing rules that prohibit repetitive rush orders and pricing wars, there may be built-in incentives for non-compliance. Moreover, even with model codes and monitoring systems in the manufacturing/processing sites, if the company's product design calls for use of raw material produced/mined exclusively in high risk parts of the world, no amount of "good" practice in the manufacturing stage could ensure the integrity of the business' efforts to eliminate child/forced labor.

3. *Sustainability*: Only practices that are integral to the whole of a business' operations can have built-in incentives for ongoing implementation. This criterion, therefore, pertains critically to the sustainability of the practice and its objectives for the business entity.

Signs / Indicators of a Holistic & Integrated Approach

- The practice should be integrated into the company's overarching commitment to *all of the core labor standards* of the ILO, in recognition of the interconnectedness of the fundamental principles and rights at work.

- Business should demonstrate that the "practice" or "intervention" is in harmony with the rest of the business' philosophy, operational principles, and practices, and integrated into the *operational guidelines* and *performance measurements* of management and employees at every level.

- Business should demonstrate that the "practice" or "intervention" includes *assessment of risks* for the use of child / forced labor *at every stage of its value*-chain – from product design, material selection, procurement, manufacturing, marketing and sales, to product teardown.
    - For example, from the product design and engineering phase, risk assessments should be conducted for the use of child/forced labor in the production of the raw materials being considered for use.

APPENDIX H

- In the case of an isolated pilot practice, there should be a clear and demonstrated commitment to scale up the practice (modified, as necessary) to be applied to other *comparable* business processes.
  - Demonstrate justifiable rationale for the choice of the particular location (or site) for the pilot.
  - Demonstrate analysis of factors that would limit replicability.
  - Demonstrate analysis of factors that enable replicability.
  - Demonstrate analysis of a "business case" for the practice, as a means to encourage replication in other comparable contexts.

- As part of the business' commitment to the sustainability of the intervention's objectives and activities (where appropriate), the practice should have integrated into its design ongoing incentives for desired behavior and decisions, and disincentives for reverting back to undesirable practices.

I. Category 2: Rooted in Community Participation & Contextual Analysis in Relation to the Target Beneficiary

Overarching Criterion: The practice should demonstrate in its problem-analysis, design, implementation, and evaluation an emphasis on (1) the *root causes* of the exploitative labor; (2) *contextual analysis*; and (3) *target community stakeholder participation*.

Among other things, this criterion has to do with the effectiveness and the *sustainability of the outcomes* of the practice. It is only when the particular practice adequately addresses the root causes of the problem and takes into account the particular socio-economic and political contexts and needs of the community, that the immediate benefits and outcomes of the practice can be sustained over time for the individual beneficiaries and the community at large. This also would contribute to prevention of further exploitative conditions for others (e.g., younger siblings and other vulnerable populations) in the community.

Signs / Indicators of Contextualized & Community-Focused Approach

- The business policies and practices in question should be in harmony with the national and local laws and policies, in so far as these standards are consistent with international standards.

- Given that forced labor and child labor occur in particular contexts with unique mix of historical, socio-economic, political, and cultural factors that enable such exploitation, the design and implementation of the practice should make evident how the particular business intervention contributes to tackling these systemic factors, addressing the root causes of the problem.

- At a minimum, the following contextual aspects should be included in the problem-analysis and intervention-design:

    Level of Government's Political Will
    - Whether the country has ratified relevant ILO conventions.
    - Whether the country Labor Code adequately addresses child labor and forced labor issues.
    - Level of government capacity for legal enforcement in the sector.

    Socio-Economic Contributors to Child/Forced Labor
    - Existence of economic (or educational) alternatives for the forced / child laborer and families.
    - Nature of the relationship between the families (and the community) and the industry sector (e.g., dependence, family tradition, etc.).
    - Community's cultural and social attitude toward child/forced labor in the sector.

    Industry Capacity & Engagement
    - Size and scope of the sector.
    - Role, engagement and level of vertical integration of industry.

- Where the root causes have been adequately analyzed and prioritized by the business, the particular practice or intervention should aim to tackle the root causes, and where appropriate, coordinate efforts with local entities, including:
    - local authorities, where appropriate
    - other businesses, where appropriate and possible
    - community leaders and members of civil society

- Where possible, the practice should include from its design the input of the intended beneficiaries, such as the local business entities, community leaders, child/forced laborers, and families of the workers.

- Where the "practice" is intended to change the behavior of other business entities along the supply chain, it should build the capacity of the local entities (e.g., small-scale farmers, local producers, etc.) for compliance.

- To the extent possible, the practice should set up adequate incentive structure for continued long-term behavioral change in the local context.

II. Category 3: Supported by Robust System of Internal & External Evaluations and Management of Data & Information

Overarching Criterion: The practice should be supported by a robust system of evaluations that assesses the effective intersection between the business' operations and the impact on the community.

APPENDIX H

Without a credible and informative system of evaluations, it would be impossible to (1) assess whether the *intended* outcomes are being reached, (2) assess what *unintended* outcomes may be associated with the practice, (3) make continual improvements on the practice for ongoing relevance, and (4) obtain information necessary for replication or scaled-up application in the sector at large.

Signs / Indicators of a Robust System of Evaluations

Evaluation Methodology

- The practice should have in its design adequate means of *internal* evaluations, which, on regular intervals:
    - Measure whether the goals are being met.
    - Assess any unintended consequences.
    - Provide a mechanism of adjusting and improving on the practices based on evaluation results.

- The practice should be supported by a set of ongoing *external* evaluations that:
    - Are conducted by entities with *specialized training* in international labor standards and in-depth knowledge of the specific commodity / product / industry. It is important that the evaluation is not a generic, cookie-cutter approach, but tailored to address the challenges specific to that sector, country, and region.
    - Prioritize the input of local community stakeholders, who demonstrate on-going, continuous presence.
    - As much as possible, include *confidential* interviews with those the "good practice" is intended to benefit / protect.
    - Demonstrate cultural and political sensitivities, including the use of the language used by the beneficiaries.
    - Move away from "pass or fail" system to creation of systemic incentives for good behavior and capacity building.
    - Utilize sampling techniques that are designed to produce high level of certainty.

- Evaluations should demonstrate capacity for effective data management:
    - Establish baseline information.
    - Give attention to establishing the reliability of the baseline data.
    - Ensure that data collection is ongoing and up-to-date.

Evaluation Components

- Evaluations should include assessment of the *logical links* in the design of the practice, including evaluation of:
    - Quality of analysis of root causes of the problem and the relationship between the intervention and the causes.
    - The logical connection between the inputs, activities, and expected outcomes.

- Assumptions underlying how the particular intervention is to lead to the desired outcomes.

◆ Evaluations should include impact measurements, which address the following questions:
    - Is the practice reducing child labor?
    - Is the practice reducing forced labor?
    - Is the practice benefiting child laborers?
    - Is the practice benefiting forced laborers?
    - What are the specific benefits?
    - How are the benefits documented?
    - How are the benefits measured?

- Evaluation should give attention to the sustainability of the outcomes and the practice (where necessary to continue prevention):
    - Demonstrate that the practice is likely to continue (as needed).
    - Demonstrate that the benefits are likely to continue for the beneficiaries.
    - Assess whether necessary institutional capacities have been built to sustain the desired outcomes.
    - Assess the degree of will that exists to ensure sustainability of the desired outcomes.
    - Assess whether there is local ownership of the practices.

At first glance, the question may arise: should all the above elements be considered necessary for a practice to be considered "good," could any corporate practice pass the test? Aren't the above criteria too onerous?

For a practice to be considered exemplary, I believe it is not only possible but necessary for a business practice to meet at least the vast majority of the three sets of criteria outlined above – if not in the letter of the criteria, at least in their spirit. It is critical that the practice be integrated into the whole of the business' operations, from design to the end of the product's life-cycle. Otherwise, whatever "good" that is intended by the particular practice may easily be countered by the forces operating in other segments of the business' value chain. Likewise, if the practice is not designed, implemented and evaluated to address the particular needs, root-causes, risks, and capacities of the local context (usually best derived at through the participation of the local community members), the practice is likely to be ineffective and unsustainable. Finally, without a robust system of evaluating the impact of the practice based on credible data management, it would be impossible for the business entity to be able to posit that the practice was effective in achieving its goals, let alone look out for the potential unintended consequences of the intervention.

There is no doubt that the Compendium that results from the application of the criteria will be watched carefully by the business community – from U.S. importers, to producers operating abroad. Whether intended or not, the practices presented in the

Compendium will be used by relevant businesses as the benchmark – the standard by which businesses will measure their own practices. As such, it is critical that the Compendium include *only* those practices that demonstrate not only reliable good results, but also good processes by which the practices were designed. I believe the three sets of criteria presented above will help point to such practices.

"Corporate social responsibility" efforts have reached a critical point in time. While much learning and improvement have taken place over the past two decades, there also have been many setbacks. The setbacks include the current prevalent system of audits and monitoring (the "policing model"), which is largely perceived as flawed and ineffective in preventing or eliminating exploitative labor conditions in major supply chains. While companies have invested considerable energy and resources to shoring up their CSR efforts, criticisms and frustrations continue, as demonstrable results are difficult to come by. In this context, it is likely that business communities will look to the U.S. Department of Labor's Compendium for guidelines on how they might shape their next round of CSR efforts. The potential role and impact of the Compendium is, in this sense, far-reaching and significant. I hope that the ideas presented in this document contribute toward building Criteria that result in a stellar set of practices that will both challenge businesses and provide useful guidelines to help shape the next phase of developments in CSR.